T0205619

Intelligent Systems Reference Library

Volume 126

Series editors

Janusz Kacprzyk, Polish Academy of Sciences, Warsaw, Poland
e-mail: kacprzyk@ibspan.waw.pl

Lakhmi C. Jain, University of Canberra, Canberra, Australia;
Bournemouth University, UK;
KES International, UK
e-mail: jainlc2002@yahoo.co.uk; jainlakhmi@gmail.com;
URL: http://www.kesinternational.org/organisation.php

Qiang Yu · Huajin Tang · Jun Hu
Kay Chen Tan

Neuromorphic Cognitive Systems

A Learning and Memory Centered Approach

 Springer

Qiang Yu
Institute for Infocomm Research
Singapore
Singapore

Jun Hu
AGI Technologies
Singapore
Singapore

Huajin Tang
College of Computer Science
Sichuan University
Chengdu
China

Kay Chen Tan
Department of Computer Science
City University of Hong Kong
Kowloon Tong
Hong Kong

ISSN 1868-4394   ISSN 1868-4408   (electronic)
Intelligent Systems Reference Library
ISBN 978-3-319-85625-4   ISBN 978-3-319-55310-8   (eBook)
DOI 10.1007/978-3-319-55310-8

Printed on acid-free paper

This Springer imprint is published by Springer Nature
The registered company is Springer International Publishing AG
The registered company address is: Gewerbestrasse 11, 6330 Cham, Switzerland

To our family, for their loves and supports.

Qiang Yu
Huajin Tang
Jun Hu
Kay Chen Tan

Preface

The powerful and yet mysterious human brain system attracts numerous researchers devoting themselves to characterizing what nervous systems do, determining how they function, and understanding why they operate in particular ways. Encompassing various studies of biology, physics, psychology, mathematics, and computer science, theoretical neuroscience provides a quantitative basis for uncovering the general principles by which the nervous systems operate. Based on these principles, neuromorphic cognitive systems introduce some basic mathematical and computational methods to describe and utilize schemes at a cognitive level. Since the mechanisms how human memory cognitively operates and how to utilize the bioinspired mechanisms to practical applications are rarely known, the study of neuromorphic cognitive systems is urgently demanded.

This book presents the computational principles underlying spike-based information processing and cognitive computation with a specific focus on learning and memory. Specifically, the action potential timing is utilized for sensory neuronal representations and computation, and spiking neurons are considered as the basic information processing unit. The topics covered in this book vary from neuronal level to system level, including neural coding, learning in both single- and multi-layered networks, cognitive memory, and applied developments of information processing systems with spiking neurons. From the neuronal level, synaptic adaptation plays an important role on learning patterns. In order to perform higher level cognitive functions such as recognition and memory, spiking neurons with learning abilities are consistently integrated with each other, building a system with the functionality of encoding, learning, and decoding. All these aspects are described with details in this book.

Theories, concepts, methods, and applications are provided to motivate researchers in this exciting and interdisciplinary area. Theoretical modeling and analysis are tightly bounded with practical applications, which would be potentially beneficial for readers in the area of neuromorphic computing. This book presents the computational ability of bioinspired systems and gives a better understanding of the mechanisms by which the nervous system might operate.

Singapore   Qiang Yu
Chengdu, China   Huajin Tang
Singapore   Jun Hu
Kowloon Tong, Hong Kong   Kay Chen Tan
December 2016

Contents

Acronyms

AMPA	Alpha-amino-3-hydroxy-5-methyl-4-isoxazolepropionic acid
ANN	Artificial neural network
CCs	Complex cells
DoG	Difference of gaussian
EPSP	Excitatory post-synaptic potential
FSM	Finite state machine
GABA	Gamma-amino-butyric-acid
GCs	Ganglion cells
HH	Hodgkin–Huxley model
IM	Izhikevich model
IPSP	Inhibitory post-synaptic potential
LIF	Leaky integrate-and-fire neuron
LSF	Local spectrogram feature
LSM	Liquid state machine
LTD	Long-term depression
LTP	Long-term potentiation
MutPSD	Multilayer PSD
MutTmptr	Multilayer tempotron
NMDA	N-methyl-D-aspartate
OCR	Optical character recognition
PSC	Post-synaptic current
PSD	Precise-spike-driven
ReSuMe	Remote supervised method
SMO	Subthreshold membrane potential oscillation
SNN	Spiking neural network
SOM	Self-organizing map
SPAN	Spike pattern association neuron
SRM	Spike response model

STDP	Spike-timing-dependent plasticity
SVM	Support vector machine
VLSI	Very large scale integration
WH	Widrow–Hoff
XOR	Exclusive OR

Chapter 1
Introduction

Abstract Since the emergence of the first digital computer, people are set free from heavy computing works. Computers can process a large amount of data with high precision and speed. However, compared to the brain, the computer still cannot approach a comparable performance considering cognitive functions such as perception, recognition and memory. For example, it is easy for human to recognize the face of a person, read papers and communicate with others, but hard for computers. Mechanisms that utilized by the brain for such powerful cognitive functions still remain unclear. Neural networks are developed for providing a brain-like information processing and cognitive computing. Theoretical analysis on neural networks could offer a key approach to revealing the secret of the brain. The subsequent sections provide detailed background information, as well as the objectives of this book.

1.1 Background

The computational power of the brain has attracted many researchers to reveal its mystery in order to understand how it works and to design human-like intelligent systems. The human brain is constructed with around 100 billion highly interconnected neurons. These neurons transmit information between each other to perform cognitive functions. Modeling neural networks facilitates investigation of information processing and cognitive computing in the brain from a mathematical point of view. Artificial neural networks (ANNs), or simply called neural networks, are the earliest work for modeling the computational ability of the brain. The research on ANNs has achieved a great deal in both theories and engineering applications. Typically, an ANN is constructed with neurons which have real-valued inputs and outputs.

However, biological neurons in the brain utilize spikes (or called as action potentials) for information transmission between each other. This phenomenon of the *spiking* nature of neurons has been known since the first experiments conducted by Adrian in the 1920s [1]. Neurons will send out short pulses of energy (spikes) as signals, if they have received enough input from other neurons. Based on this mechanism, spiking neurons are developed with a same capability of processing spikes as biological neurons. Thus, spiking neural networks (SNNs) are more biologically

© Springer International Publishing AG 2017
Q. Yu et al., *Neuromorphic Cognitive Systems*, Intelligent Systems
Reference Library 126, DOI 10.1007/978-3-319-55310-8_1

plausible than ANNs since the concept of spikes, rather than real values, is considered in the computation. SNNs are widely studied in recent years, but questions of how information is represented by spikes and how the neurons process these spikes are still unclear. These two questions demand further studies on neural coding and learning in SNNs.

Spikes are believed to be the principal feature in the information processing of neural systems, though the neural coding mechanism remains unclear. In 1920s, Adrian also found that sensory neurons fire spikes at a rate monotonically increasing with the intensity of stimulus. This observation led to the widespread adoption of the hypothesis of a rate coding, where neurons communicate purely through their firing rates. Recently, an increasing body of evidence shows that the precise timing of individual spikes also plays an important role [2]. This finding supports the hypothesis of a temporal coding, where the precise timing of spikes, rather than the rate, is used for encoding information. Within a *temporal coding* framework, temporal learning describes how neurons process precise-timing spikes. Further research on temporal coding and temporal learning would provide a better understanding of the biological systems, and would also explore potential abilities of SNNs for information processing and cognitive computing. Moreover, beyond independently studying the temporal coding and learning, it would be more important and useful to consider both in a consistent system.

1.2 Spiking Neurons

The rough concept of how neurons work is understood: neurons send out short pulses of electrical energy as signals, if they have received enough of these themselves. This principal mechanism has been modeled into various mathematical models for computer use. These models are built under the inspiration of how real neurons work in the brain.

1.2.1 Biological Background

A neuron is an electrically excitable cell that processes and transmits information by electrical and chemical signaling. Chemical signaling occurs via synapses, specialized connections with other cells. Neurons form neural networks through connecting with each other.

Computers communicate with bits; neurons use spikes. Incoming signals change the membrane potential of the neuron and when it reaches above a certain value the neuron sends out an action potential (spike).

As is shown in Fig. 1.1, a typical neuron possesses a cell body (often called soma), dendrites, and an axon. The dendrites serve as the inputs of the neuron and the axon

Fig. 1.1 Structure of a
typical neuron. A neuron
typically possesses a soma,
dendrites and an axon. The
neuron receives inputs via
dendrites and sends output
through the axon

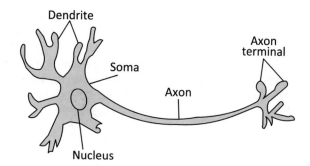

acts as the output. The neuron collects information through its dendrites and sends
out the reaction through the axon.

Spikes cannot cross the gap between one neuron and the other. Connections
between neurons are formed via cellular interfaces, so called synapses. An incoming
pre-synaptic action potential triggers the release of neurotransmitter chemicals in
vesicles. These neurotransmitters cross the synaptic gap and bind to receptors on the
dendritic side of the synapse. Then a post-synaptic potential will be generated [3, 4].

The type of synapse and the amount of released neurotransmitter determine the
type and strength of the post-synaptic potential. The membrane potential would be
increased by excitatory post-synaptic potential (EPSP) or decreased by inhibitory
post-synaptic potential (IPSP). Real neurons only use one type of neurotransmitter in all their outgoing synapses. This makes the neuron either be excitatory or
inhibitory [3].

1.2.2   Generations of Neuron Models

From the conceptual point of view, all neuron models share the following common
features:

1. **Multiple inputs and single output**: The neuron receives many inputs and produces a single output signal.
2. **Different types of inputs**: The output activities of neurons are characterized by
   at least one state variable that usually corresponding to the membrane potential. An input from the excitatory/inhibitory synapses will increase/decrease the
   membrane potential.

Based on these conceptual features, various neuron models are developed. Artificial neural networks are already becoming a fairly old technique within computer
science. The first ideas and models are over fifty years old. The first generation of
artificial neuron is the one with McCulloch-Pitts threshold. These neurons can only
give digital output. Neurons of the second generation do not use a threshold function to compute their output signals, but a continuous activation function, making

them suitable for analog input and output [5]. Typical examples of neural networks consisting of these neurons are feedforward and recurrent neural networks. They are more powerful than their first generation [6].

Neuron models of the first two generations do not employ the individual pulses. The third generation of neuron models raises the level of biological realism by using individual spikes. This allows incorporating spatiotemporal information in communication and computation, like real neurons do.

1.2.3 Spiking Neuron Models

For the reasons of greater computational power and more biological plausibility, spiking neurons are widely studied in recent years. As the third generation of neuron models, spiking neurons increase the level of realism in a neural simulation.

Spiking neurons have an inherent notion of time that makes them seemingly particularly suited for processing temporal input data [7]. Their nonlinear reaction to input provides them with strong computational qualities, theoretically requiring just small networks for complex tasks.

1.2.3.1 Leaky Integrate-and-Fire Neuron (LIF)

The leaky integrate-and-fire neuron [4] is the most widely used and best-known model of threshold-fire neurons. The membrane potential of the neuron $V_m(t)$ is dynamically changing over time, as:

$$\tau_m \frac{dV_m}{dt} = -V_m + I(t) \tag{1.1}$$

where τ_m is the membrane time constant in which voltage *leaks* away. A bigger τ_m can result in a slower decaying process of $V_m(t)$. $I(t)$ is the input current which is a weighted sum from all incoming spikes.

Once a spike arrives, it is multiplied by corresponding synaptic efficacy factor to form the post-synaptic potential that changes the potential of the neuron. When the membrane potential crosses a certain threshold value, the neuron will elicit a spike; after which the membrane potential goes back to a reset value and holds there for a refractory period. Within the refractory time, the neuron is not allowed to fire.

From both the conceptual and computational points of view, the LIF model is relatively simple comparing to other spiking neuron models. An advantage of the model is that it is relatively easy to integrate it in hardware, achieving a very fast operation. Various generalizations of the LIF model have been developed. One popular generalization of the LIF model is the Spike Response Model (SRM), where a kernel approach is used in neuron's dynamics. The SRM is widely used due to its simplicity in analysis.

1.2.3.2 Hodgkin-Huxley Model (HH) and Izhikevich Model (IM)

The Hodgkin-Huxley (HH) model was based on experimental observations with the large neurons found in squid [8]. It is by far the most detailed and complex neuron model. However, this model is less suited for simulations of large networks since the realism of neuron model comes at a large computational cost.

The Izhikevich model (IM) was proposed in [9]. By choosing different parameter values in the dynamic equations, the neuron model can function differently, like bursting or single spiking.

1.3 Neural Codes

The world around us is extremely dynamic, that everything changes continuously over time. The information of the external world goes into our brain through the sensory systems. Determining how neuronal activity represents sensory information is central for understanding perception. Besides, understanding the representation of external stimuli in the brain directly determines what kind of information mechanism should be utilized in the neural network.

Neurons are remarkable among the cells of the body in their ability to propagate signals rapidly over large distances. They do this by generating characteristic electrical pulses called action potentials or, more simply, spikes that can travel down nerve fibers. Sensory neurons change their activities by firing sequences of action potentials in various temporal patterns, with the presence of external sensory stimuli, such as light, sound, taste, smell and touch. It is known that information about the stimulus is encoded in this pattern of action potentials and transmitted into and around the brain.

Although action potentials can vary somewhat in duration, amplitude and shape, they are typically treated as identical stereotyped events in neural coding studies. Action potentials are all very similar. In addition, neurons in the brain work together, rather than individually, to transfer the information.

Figure 1.2 shows a typical spatiotemporal spike pattern. This pattern contains both spatial and temporal information of a neuron group. Each neuron fires a spike train within a time period. The spike trains of the whole neuron group form the spatiotemporal pattern. The spiking neurons inherently aim to process and produce this kind of spatiotemporal spike patterns.

The question is still not clear that how this kind of spike trains could convey information of the external stimuli. A spike train may contain information based on different coding schemes. In motor neurons, for example, the strength at which an innervated muscle is flexed depends solely on the 'firing rate', the average number of spikes per unit time (a 'rate code'). At the other end, a complex 'temporal code' is based on the precise timing of single spikes. They may be locked to an external stimulus such as in the auditory system or be generated intrinsically by the neural circuitry [10].

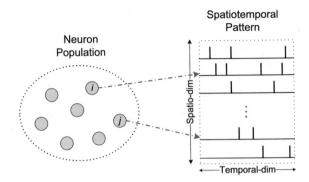

Fig. 1.2 A typical spatiotemporal spike pattern. A group of neurons (Neuron Group) works together to transfer the information, with each neuron firing a spike train in time. All spike trains from the group form a pattern with both spatio- and temporal-dimension information. This is called spatiotemporal spike pattern. The vertical lines denote spikes

Whether neurons use the rate code or the temporal code is a topic of intense debate within the neuroscience community, even though there is no clear definition of what these terms mean. The followings further present a detailed overview of the rate code and the temporal code.

1.3.1   Rate Code

Rate code is a traditional coding scheme, assuming that most, if not all, information about the stimulus is contained in the firing rate of the neuron. Because the sequence of action potentials generated by a given stimulus varies from trial to trial, neuronal responses are treated statistically or probabilistically. They may be characterized by firing rates, rather than by specific spike sequences. In most sensory systems, the firing rate increases, generally non-linearly, with increasing stimulus intensity [3]. Any information possibly encoded in the temporal structure of the spike train is ignored. Consequently, the rate code is inefficient but highly robust with respect to input noise.

Before encoding external information into firing rates, precise calculation of the firing rates is required. In fact, the term 'firing rate' has a few different definitions, which refer to different averaging procedures, such as an average over time or an average over several repetitions of experiment. For most cases in the coding scheme, it considers the spike count within an encoding window [11]. The encoding window is defined as the temporal window that contains the response patterns that are considered as the basic information-carrying units of the code. The hypothesis of the rate code receives support from the ubiquitous correlation of firing rates with sensory variables [1].

1.3.2 Temporal Code

When precise spike timing or high-frequency firing-rate fluctuations are found to carry information, the neural code is often identified as a temporal code [12]. A number of studies have found that the temporal resolution of the neural code is on a millisecond time scale, indicating that precise spike timing is a significant element in neural coding [13, 14].

Neurons, in the retina [15, 16], the lateral geniculate nucleus (LGN) [17] and the visual cortex [14, 18] as well as in many other sensory systems [19, 20], are observed to precisely respond to the stimulus on a millisecond timescale. These experiments support hypothesis of the temporal code, in which precise timings of spikes are taken into account for conveying information.

Like real neurons, communication is based on individually timed pulses. The temporal code is potentially much more powerful for encoding information with respect to the rate code. It is possible to multiplex much more information into a single stream of individual pulses than you can transmit using just the average firing rates of a neuron. For example, the auditory system can combine the information of amplitude and frequency very efficiently over one single channel [21].

Another advantage of the temporal code is speed. Neurons can be made to react to single spikes, allowing for extremely fast binary calculation. The human brain, for example, can recognize faces in as little as 100 ms [22, 23].

There are several kinds of temporal code that have been proposed, like latency code, interspike intervals code and phase of firing code [11]. Latency code is a specific form of temporal code, that encoding information in the timing of response relative to the encoding window, which is usually defined with respect to stimulus onset. The latency of a spike is determined by the external stimuli. A stronger input could result in an earlier spike. In the interspike intervals code, the temporally encoded information is carried by the relative time between spikes, rather than by their absolute time with respect to stimulus onset. In the phase of firing code, information is encoded by the relative timing of spikes regarding to the phase of subthreshold membrane oscillations [11, 24].

1.3.3 Temporal Code Versus Rate Code

In the rate code, a higher sensory variable corresponds to a higher firing rate. Although there are few doubts as to the relevance of this firing rate code, it neglects the extra information embedded in the temporal structure.

Recent studies have shown neurons in the vertebrate retina fire with remarkable temporal precision. In addition, temporal patterns in spatiotemporal spikes can carry more information than the rate-based code [25–27]. Thus, temporal code serves as an important component in neural system.

Since the temporal code is more biologically plausible and computationally powerful, a temporal framework is considered throughout this study.

1.4 Cognitive Learning and Memory in the Brain

In a biological nervous system, learning and memory are two indispensable components of all mental processes, especially for cognition functions. Learning is the acquisition of new knowledge or modification of existing knowledge through study and experience. And memory is the process in which information is encoded, stored, and retrieved. Therefore, an intelligent machine is supposed to be able to obtain knowledge from external stimulation and store them in the form of memory. When encountering new problems, it would response relying on the stored knowledge.

1.4.1 Temporal Learning

Researchers have gone a long way to explore the secret of learning mechanisms in the brain. In neuroscience, the learning process is found to be related to synaptic plasticity, where the synaptic weights are adjusted along the learning.

In 1949, Donald Hebb introduced a basic mechanism that explained the adaptation of neurons in the brain during the learning process [28]. It is called the Hebbian learning rule, where a change in the strength of a connection is a function of the pre- and post-synaptic neural activities. When neuron A repeatedly participates in firing neuron B, the synaptic weight from A to B will be increased.

The Hebbian mechanism has been the primary basis for learning rules in spiking neural networks, though detailed processes of the learning occurring in biological systems are still unclear. According to the schemes on how information is encoded with spikes, learning rules in spiking neural networks can be generally assorted into two categories: rate learning and temporal learning.

The rate learning algorithms, such as the spike-driven synaptic plasticity rule [29, 30], are developed for processing spikes presented in a rate-based framework, where mean firing rates of the spikes are used for carrying information. However, since the rate learning algorithms are formulated in a rate-based framework, this group of rules cannot be applied to process precise-time spike patterns.

To process spatiotemporal spike patterns with a temporal framework, the temporal learning rule is developed. This kind of learning rule can be used to process information that is encoded with a temporal code, where precise timing of spikes acts as the information carrier. Development of the temporal learning rule is imperative considering an increasing body of evidence supporting the temporal code.

Among various temporal rules, several rules have been widely studied, including: spike-timing-dependent plasticity (STDP) [31, 32], the tempotron rule [33], the

SpikeProp rule [34], the SPAN rule [35], the Chronotron rule [36] and the ReSuMe rule [37].

STDP is one of the most commonly and experimentally studied rules in recent years. STDP is in agreement with Hebbs postulate because it reinforces the connections with the pre-synaptic neurons that fired slightly before the postsynaptic neuron, which are those that *took part in firing it*. STDP describes the learning process depending on the precise spike timing, which is more biologically plausible. The STDP modification rule is shown as the following equation:

$$\Delta w_{ij} = \begin{cases} A^+ \cdot exp(\dfrac{\Delta t}{\tau^+}) & : \Delta t \leq 0 \\ -A^- \cdot exp(-\dfrac{\Delta t}{\tau^-}) & : \Delta t > 0 \end{cases} \tag{1.2}$$

where Δt denotes the time difference between the pre- and post-synaptic spikes. A^+, A^- and τ^+, τ^- are parameters of learning rates and time constants, respectively.

As is shown in Fig. 1.3a, if pre-synaptic spike fire before the post-synpatic spike, long-term potentiation (LTP) will happen. Long-term depression (LTD) occurs when the firing order is reversed.

Figure 1.3b shows that neurons equipped with STDP can automatically find the repeating pattern which is embedded in a background. The neuron will emit a spike at the presence of this pattern [38–40].

However, STDP characterizes synaptic changes solely in terms of the temporal contiguity of the pre-synaptic spike and the post-synaptic potential or spike. This is not enough for learning spatiotemporal patterns since it would cause silent response sometimes.

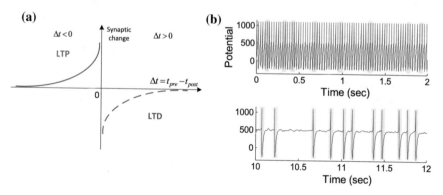

Fig. 1.3 Spike-Timing-Dependent Plasticity (STDP). **a** is a typical asymmetric learning window of STDP. Pre-synaptic spike firing before post-synaptic spike will cause long-term potentiation (LTP). Long-term depression (LTD) occurs if the order of these two spikes is reversed. **b** shows the ability of STDP to learn and detect repeating patterns that embedded in continuous spike trains. Shaded areas denote the embedded repeating patterns, and the blue line shows the potential trace of the neuron. Along the learning with STDP, the neuron gradually detects the target pattern by firing a spike

The tempotron rule [33] is one such temporal learning rule where neurons are trained to discriminate between two classes of spatiotemporal patterns. This learning rule is based on a gradient descent approach. In the tempotron rule, the synaptic plasticity is governed by the temporal contiguity of pre-synaptic spike, post-synaptic depolarization and a supervisory signal. The neurons could be trained to successfully distinguish two classes by firing a spike or by remaining quiescent.

The tempotron rule is an efficient rule for the classification of spatiotemporal patterns. However, the neurons do not learn to fire at precise time. Since the tempotron rule mainly aims at decision-making tasks, it cannot support the same coding scheme used in both the input and output spikes. The time of the output spike seems to be arbitrary, and does not carry information. To support the same coding scheme through the input and output, a learning rule is needed to let the neuron not only fire but also fire at the specified time. In addition, the tempotron rule is designed for a specific neuron model, which might limit its usage on other spiking neuron models.

By contrast, the SpikeProp rule [34] can train neurons to perform a spatiotemporal classification by emitting single spikes at the desired firing time. The SpikeProp rule is a supervised learning rule for SNNs that based on gradient descent approach. The major limitation is that the SpikeProp rule and its extension in [41] do not allow multiple spikes in the output spike train. To solve this problem, several other temporal learning rules, such as the SPAN rule, the Chronotron rule and the ReSuMe rule, have been developed to train neurons to produce multiple output spikes in response to a spatiotemporal stimulus.

In both the SPAN rule and the Chronotron E-learning rule, the synaptic weights are modified according to a gradient descent approach in an error landscape. The error function in the Chronotron rule is based on the Victor and Purpura distance [42] in which the distance between two spike trains is defined as the minimum cost required to transform one into the other, while in the SPAN rule the error function is based on a metric similar to the van Rossum metric [43] where spike trains are converted into continuous time series for evaluating the difference. These arithmetic calculations can easily reveal why and how networks with spiking neurons can be trained, but the arithmetic-based rules are not a good choice for designing networks with biological plausibility. The biological plausibility of error calculation is at least questionable.

From the perspective of increased biological plausibility, the Chronotron I-learning rule and the ReSuMe rule are considered. The I-learning rule isheuristically defined in [36] where synaptic changes depend on the synaptic currents. According to the I-learning rule, its development seems to be based on a particular case of the Spike Response Model [4], which might also limit its usage on other spiking neuron models or at least is not clearly demonstrated. Moreover, those synapses with zero initial weights will never be updated according to the I-learning rule. This will inevitably lead to information loss from those afferent neurons.

In view of the two aspects presented above, i.e., the biological plausibility and the computational efficiency, it is important to combine the two aspects for a new temporal learning rule and develop a comprehensive research framework within a system where information is carried by precise-timing spikes.

1.4.2 Cognitive Memory in the Brain

From the perspective of psychology, memory can be generally divided into sensory memory, short-term memory (STM) and long term memory (LTM). Memory models proposed by psychologists provide abstract descriptions of how memory works. For example, Atkinson-Shiffrin model simplifies the memory system as shown in Fig. 1.4.

In order to explore the memory function in biological systems, different parts of biological nervous system have been studied [45, 46]. Researches on sensory system, especially vision system, advance our understanding of sensory encoding. Indicated by the study on patient H.M., the hippocampus is believed to be the most essential part involved in consolidation of memory.

Fig. 1.4 Multi-store model of Memory [44]

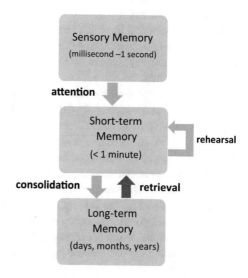

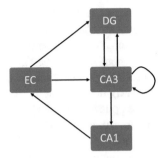

Fig. 1.5 Diagram of the hippocampal circuit. Entorhinal cortex (EC), dentate gyrus (DG) and Cornu Ammonis areas CA3 and CA1 are the main components of hippocampus

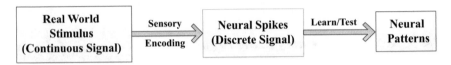

Fig. 1.6 Information flow in a typical neural network. Analogue stimulus is encoded by sensory organs into neural signals and propagate in the neural network

The hippocampus is one of the most widely studied regions of the brain. It resides within the medial temporal lobe of the brain and is believed to be responsible for the storage of declarative memories. At the macroscopic level, highly processed neocortical information from all sensory inputs converges onto the medial temporal lobe. These processed signals enter the hippocampus via the entorhinal cortex (EC). Within the hippocampus, there are connections from the EC to all parts of the hippocampus, including the dentate gyrus (DG), Cornu Ammonis areas CA3 and CA1 through perforant pathway. Connections from the DG are connected to CA3 through mossy fibers, from CA3 to CA1 through schaffer collaterals, and then from CA1 back to EC. In addition, there are also strong recurrent connections within the DG and CA3 regions. Figure 1.5 depicts the overview of hippocampus based on its anatomic structure.

A few computational models simulating different regions of the hippocampus were proposed and studied [47–49]. Inspired by the structure of hippocampus, memory function have been demonstrated in these models. However, due to insufficient knowledge about mechanisms underlying neural computation in biological systems, limited function of the brain can be artificially reproduced with current technology. By simulating artificial neuron models, neural networks are inherently close to the nature of biological nervous systems and possible to mimic their functions. Figure 1.6 illustrates the information flow in typical artificial neural networks. Similar to biological nervous systems, encoding and learning are the most important components of a memory model using neural networks. Encoding defines the form of neural representation of information, while learning plays a pivotal role in the development of neural systems and formation of memory.

Due to their more biological realistic properties, spiking neural networks have enhanced our understanding of information processing in the biological system and advanced research of computational neuroscience over the past few decades. Moreover, increasing experimental findings in neurobiology and research achievements in computational intelligence using spiking neural networks lead to growing research interests in designing learning and memory models with spiking neurons.

1.5 Objectives and Contributions

Even though many attempts have been devoted to exploring mechanisms used in the brain, a majority of facts about spiking neurons for information processing and cognitive computing still remain unclear. The research gaps for current studies on SNNs are summarized below:

1. Temporal coding and temporal learning are two of the major areas in SNNs. Various mechanisms are proposed based on inspirations from biological observations. However, most studies on these two areas are independent. There are few studies considering both the coding and the learning in a consistent system [30, 34, 50–52].
2. Over the rate-based learning algorithms, the temporal learning algorithms are developed for processing precise-timing spikes. However, these temporal learning algorithms focus more on the aspects of either arithmetic or biological plausibility. Either side of these two aspects would not be a good choice considering both the computational efficiency and the biological plausibility.
3. Currently, there are few studies on the practical applications of SNNs [30, 34, 51–53]. Most studies only focus on theoretical explorations of SNNs.
4. Learning mechanisms for building causal connections have not been clearly investigated.
5. Research on memory models carried out so far mainly focuses on specific regions of the brain, ignoring the underlying memory organization principle at a network level.

This book mainly focuses on how to explore and develop cognitive computations with spiking neurons under a temporal framework. The specific objectives covered are as follows:

1. To develop integrated consistent systems of spiking neurons, where both the coding and the learning are considered from a systematic level.
2. To develop a new temporal learning algorithm that is both simple for computation and also biologically plausible.
3. To investigate various properties of the proposed algorithm, such as memory capacity, robustness to noise and generality to different neuron models, etc.

4. To investigate the ability of the proposed SNNs applying to different cognitive tasks, such as image recognition, sound recognition and sequence recognition, etc.
5. To investigate the temporal learning in multilayer spiking neural networks.
6. To develop computational neural network models emphasizing the generation and organization of memory.

The significance of this book is two-fold. On one hand, such models introduced in the book may contribute to a better understanding of the mechanisms by which the real brains operate. On the other hand, the computational models inspired from biology are interesting in their own right, and could provide meaningful techniques for developing real-world applications.

The computations of spiking neurons in this book are considered in a temporal framework rather than a rate-based framework. This is because mounting evidence shows that precise timing of individual spikes plays an important role. In addition, the temporal framework offers significant computational advantages than the rate-based framework.

1.6 Outline of the Book

In the area of theoretical neuroscience, the general target is to provide a quantitative basis for describing what nervous systems do, understanding how they function, and uncovering the general principles by which they operate. It is a challenging area since multidisciplinary knowledges are required for building models. Investigating spike-based computation serves as a main focus in this book. To further specify the research scope, the temporal framework is considered in this book. In order to achieve the aforementioned objectives, a general system structure has been devised. Further investigations on individual functional parts of the system are conducted. The organization of the remaining chapters of this book is as follows.

Chapter 2 presents a brain-inspired spiking neural network system with simple temporal encoding and learning. With a biologically plausible supervised learning rule, the system is applied to various pattern recognition tasks. The proposed approach is also benchmarked with the nonlinearly separable task. The encoding system provides different levels of robustness, and enables the spiking neural networks to process real-world stimuli, such as images and sounds. Detailed investigations on the encoding and learning are also provided.

Chapter 3 introduces a computational model with spike-timing-based encoding schemes and learning algorithms in order to bridge the gap between sensory encoding and synaptic information processing. By treating sensory coding and learning as a systematic process, the integrated model performs sensory neural encoding and supervised learning with precisely timed spikes. We show that with a supervised spike-timing-based learning, different spatiotemporal patterns can be recognized by reproducing spike trains with a high time precision in milliseconds.

In Chap. 4, a novel learning rule, namely Precise-Spike-Driven (PSD) synaptic plasticity, is proposed for training the neuron to associate spatiotemporal spike patterns. The PSD rule is simple, efficient, and biologically plausible. Various properties of this rule are investigated.

Chapter 5 presents the application of the PSD rule on sequence recognition. In addition, the classification ability of the PSD rule is investigated and benchmarked against other learning rules.

In Chap. 6, the learning in multilayer spiking neural networks is investigated. Causal connections are built to facilitate the learning. Several tasks are used to analyze the learning performance of the multilayer network.

Chapter 7 describes a hierarchically organized memory model using spiking neurons, in which temporal population codes are considered as the neural representation of information and spike-timing-based learning methods are employed to train the network. It has been demonstrated that neural cliques representing patterns are generated and input patterns are stored in the form of associative memory within gamma cycles. Moreover, temporally separated patterns can be linked and compressed via enhanced connections among neural groups forming episodic memory.

Finally, Chap. 8 presents spiking neuron based cognitive memory models.

References

1. Adrian, E.: The Basis of Sensation: The Action of the Sense Organs. W. W. Norton, New York (1928)
2. VanRullen, R., Guyonneau, R., Thorpe, S.J.: Spike times make sense. Trends Neurosci. **28**(1), 1–4 (2005)
3. Kandel, E.R., Schwartz, J.H., Jessell, T.M., et al.: Principles of Neural Science, vol. 4. McGraw-Hill, New York (2000)
4. Gerstner, W., Kistler, W.M.: Spiking Neuron Models: Single Neurons, Populations, Plasticity, 1st edn. Cambridge University Press, Cambridge (2002)
5. Vreeken, J.: Spiking neural networks, an introduction. Institute for Information and Computing Sciences, Utrecht University Technical Report UU-CS-2003-008 (2002)
6. Maass, W., Schnitger, G., Sontag, E.D.: On the computational power of sigmoid versus boolean threshold circuits. In: 32nd Annual Symposium on Foundations of Computer Science, 1991. Proceedings, pp. 767–776. IEEE (1991)
7. Hopfield, J.J., Brody, C.D.: What is a moment? transient synchrony as a collective mechanism for spatiotemporal integration. Proc. Natl. Acad. Sci. **98**(3), 1282–1287 (2001)
8. Hodgkin, A.L., Huxley, A.F.: A quantitative description of membrane current and its application to conduction and excitation in nerve. J. Physiol. **117**(4), 500–544 (1952)
9. Izhikevich, E.M.: Simple model of spiking neurons. IEEE Trans. Neural Netw. **14**(6), 1569–1572 (2003)
10. Gerstner, W., Kreiter, A.K., Markram, H., Herz, A.V.: Neural codes: firing rates and beyond. Proc. Natl. Acad. Sci. **94**(24), 12740–12741 (1997)
11. Panzeri, S., Brunel, N., Logothetis, N.K., Kayser, C.: Sensory neural codes using multiplexed temporal scales. Trends Neurosci. **33**(3), 111–120 (2010)
12. Dayan, P., Abbott, L.: Theoretical neuroscience: computational and mathematical modeling of neural systems. J. Cognit. Neurosci. **15**(1), 154–155 (2003)

13. Butts, D.A., Weng, C., Jin, J., Yeh, C.I., Lesica, N.A., Alonso, J.M., Stanley, G.B.: Temporal precision in the neural code and the timescales of natural vision. Nature **449**(7158), 92–95 (2007)
14. Bair, W., Koch, C.: Temporal precision of spike trains in extrastriate cortex of the behaving macaque monkey. Neural Comput. **8**(6), 1185–1202 (1996)
15. Berry, M.J., Meister, M.: Refractoriness and neural precision. J. Neurosci. **18**(6), 2200–2211 (1998)
16. Uzzell, V.J., Chichilnisky, E.J.: Precision of spike trains in primate retinal ganglion cells. J. Neurophysiol. **92**(2), 780–789 (2004)
17. Reinagel, P., Reid, R.C.: Temporal coding of visual information in the thalamus. J. Neurophysiol. **20**(14), 5392–5400 (2000)
18. Mainen, Z., Sejnowski, T.: Reliability of spike timing in neocortical neurons. Science **268**(5216), 1503–1506 (1995)
19. Gabbiani, F., Metzner, W., Wessel, R., Koch, C.: From stimulus encoding to feature extraction in weakly electric fish. Nature **384**(6609), 564–567 (1996)
20. Wehr, M., Zador, A.M.: Balanced inhibition underlies tuning and sharpens spike timing in auditory cortex. Nature **426**(6965), 442–446 (2003)
21. Maass, W., Bishop, C.M.: Pulsed Neural Networks. MIT Press (2001)
22. Serre, T., Oliva, A., Poggio, T.: A feedforward architecture accounts for rapid categorization. Proc. Natl. Acad. Sci. **104**(15), 6424–6429 (2007)
23. Gollisch, T., Meister, M.: Rapid neural coding in the retina with relative spike latencies. Science **319**(5866), 1108–1111 (2008)
24. Nadasdy, Z.: Information encoding and reconstruction from the phase of action potentials. Front. Syst. Neurosci. **3**, 6 (2009)
25. Kempter, R., Gerstner, W., van Hemmen, J.L.: Spike-based compared to rate-based Hebbian learning. In: NIPS'98, pp. 125–131 (1998)
26. Borst, A., Theunissen, F.E.: Information theory and neural coding. Nat. Neurosci. **2**(11), 947–957 (1999)
27. Hopfield, J.J.: Pattern recognition computation using action potential timing for stimulus representation. Nature **376**(6535), 33–36 (1995)
28. Hebb, D.: The Organization of Behavior: A Neuropsychological Theory. Taylor & Francis Group (2002)
29. Fusi, S.: Spike-driven synaptic plasticity for learning correlated patterns of mean firing rates. Rev. Neurosci. **14**(1–2), 73–84 (2003)
30. Brader, J.M., Senn, W., Fusi, S.: Learning real-world stimuli in a neural network with spike-driven synaptic dynamics. Neural Comput. **19**(11), 2881–2912 (2007)
31. Bi, G.Q., Poo, M.M.: Synaptic modification by correlated activity: Hebb's postulate revisited. Ann. Rev. Neurosci. **24**, 139–166 (2001)
32. Froemke, R.C., Poo, M.M., Dan, Y.: Spike-timing-dependent synaptic plasticity depends on dendritic location. Nature **434**(7030), 221–225 (2005)
33. Gütig, R., Sompolinsky, H.: The tempotron: a neuron that learns spike timing-based decisions. Nat. Neurosci. **9**(3), 420–428 (2006)
34. Bohte, S.M., Kok, J.N., Poutré, J.A.L.: Error-backpropagation in temporally encoded networks of spiking neurons. Neurocomputing **48**(1–4), 17–37 (2002)
35. Mohemmed, A., Schliebs, S., Matsuda, S., Kasabov, N.: SPAN: spike pattern association neuron for learning spatio-temporal spike patterns. Int. J. Neural Syst. **22**(04), 1250,012 (2012)
36. Florian, R.V.: The Chronotron: a neuron that learns to fire temporally precise spike patterns. PLoS One **7**(8), e40,233 (2012)
37. Ponulak, F.: ReSuMe-new supervised learning method for spiking neural networks. Institute of Control and Information Engineering, Poznoń University of Technology, Tech. rep. (2005)
38. Guyonneau, R., van Rullen, R., Thorpe, S.J.: Neurons tune to the earliest spikes through STDP. Neural Comput. **17**(4), 859–879 (2005)
39. Masquelier, T., Guyonneau, R., Thorpe, S.J.: Spike timing dependent plasticity finds the start of repeating patterns in continuous spike trains. PloS One **3**(1), e1377 (2008)

40. Masquelier, T., Guyonneau, R., Thorpe, S.J.: Competitive stdp-based spike pattern learning. Neural Comput. **21**(5), 1259–1276 (2009)
41. Booij, O., et al.: A gradient descent rule for spiking neurons emitting multiple spikes. Inf. Process. Lett. **95**(6), 552–558 (2005)
42. Victor, J.D., Purpura, K.P.: Metric-space analysis of spike trains: theory, algorithms and application. Netw.: Comput. Neural Syst. **8**(2), 127–164 (1997)
43. Van Rossum, M.C., Bi, G., Turrigiano, G.: Stable Hebbian learning from spike timing-dependent plasticity. J. Neurosci. **20**(23), 8812–8821 (2000)
44. Atkinson, R.C., Shiffrin, R.M.: Human memory: A proposed system and its control processes. In: Spence, K.W., Spence, J.T. (eds.) The Psychology of Learning and Motivation: Advances in Research and Theory, vol. 2, pp. 89–105 (1968)
45. Hawkins, J., Blakeslee, S.: On Intelligence. Macmillan (2004)
46. He, H.: Self-adaptive Systems for Machine Intelligence. Wiley (2011)
47. Jensen, O., Lisman, J.E.: Hippocampal sequence-encoding driven by a cortical multi-item working memory buffer. Trends Neurosci. **28**(2), 67–72 (2005)
48. Kunec, S., Hasselmo, M.E., Kopell, N.: Encoding and retrieval in the ca3 region of the hippocampus: a model of theta-phase separation. J. Neurophysiol. **94**(1), 70–82 (2005)
49. Cutsuridis, V., Wennekers, T.: Hippocampus, microcircuits and associative memory. Neural Netw. **22**(8), 1120–1128 (2009)
50. Dennis, J., Yu, Q., Tang, H., Tran, H.D., Li, H.: Temporal coding of local spectrogram features for robust sound recognition. In: 2013 IEEE International Conference on Acoustics, Speech and Signal Processing (ICASSP), pp. 803–807 (2013)
51. Yu, Q., Tang, H., Tan, K.C., Li, H.: Rapid feedforward computation by temporal encoding and learning with spiking neurons. IEEE Trans. Neural Netw. Learn. Syst. **24**(10), 1539–1552 (2013)
52. Wade, J.J., McDaid, L.J., Santos, J.A., Sayers, H.M.: SWAT: a spiking neural network training algorithm for classification problems. IEEE Trans. Neural Netw. **21**(11), 1817–1830 (2010)
53. Ponulak, F., Kasinski, A.: Supervised learning in spiking neural networks with resume: sequence learning, classification, and spike shifting. Neural Comput. **22**(2), 467–510 (2010)

Chapter 2
Rapid Feedforward Computation by Temporal Encoding and Learning with Spiking Neurons

Abstract As we know, primates perform remarkably well in cognitive tasks such as pattern recognition. Motivated from recent findings in biological systems, a unified and consistent feedforward system network with a proper encoding scheme and supervised temporal rules is built for processing real-world stimuli. The temporal rules are used for processing the spatiotemporal patterns. To utilize these rules on images or sounds, a proper encoding method and a unified computational model with consistent and efficient learning rule are required. Through encoding, external stimuli are converted into sparse representations which also have properties of invariance. These temporal patterns are then learned through biologically derived algorithms in the learning layer, followed by the final decision presented through the readout layer. The performance of the model is also analyzed and discussed. This chapter presents a general structure of SNN for pattern recognition, showing that the SNN has the ability to learn the real-world stimuli.

2.1 Introduction

Primates are remarkably good at cognitive skills such as pattern recognition. Despite decades of engineering effort, the performance of the biological visual system still outperforms the best computer vision systems. Pattern recognition is a general task that assigns an output value to a given input pattern. It is an information-reduction process which aims to classify patterns based on a priori knowledge or statistical information extracted from the patterns. Typical applications of pattern recognition includes automatic speech recognition, handwritten postal codes recognition, face recognition and gesture recognition. There are several conventional methods to implement pattern recognition, such as maximum entropy classifier, Naive Bayes classifier, decision trees, support vector machines (SVM) and perceptrons. We refer these methods as traditional rules since they are less biologically plausible compared to spiking time involved rules described later. Compared to human brain, these methods are far from reaching comparable recognition. Humans can easily discriminate different categories within a very short time. This motivates us to investigate computational models for rapid and robust pattern recognition from a biological point of

© Springer International Publishing AG 2017

Q. Yu et al., *Neuromorphic Cognitive Systems*, Intelligent Systems
Reference Library 126, DOI 10.1007/978-3-319-55310-8_2

view. At the same time, inspired by biological findings, researchers have come up with different theories of encoding and learning. In order to bridge the gap between those independent studies, a unified systematic model with consistent rules is desired.

A simple feedforward architecture might account for rapid recognition as reported recently [1]. Anatomical back projections abundantly appear almost every area in the visual cortex, which puts the feedforward architecture under debate. However, the observation of a quick response appeared in inferotemporal cortex (IT) [2] most directly supports the hypothesis of the feedforward structure. The activity of neurons in monkey IT appears quite soon (around 100 ms) after stimulus onset [3]. For the purpose of rapid recognition, a core feedforward architecture might be a reasonable theory of visual computation.

How information is represented in the brain still remains unclear. However, there are strong reasons to believe that using pulses is the optimal way to encode the information for transmission [4]. Increasing number of observations show that neurons in the brain precisely response to a stimulus. This support the hypothesis of the temporal coding.

There are many temporal learning rules proposed for processing spatiotemporal patterns, including both supervised and unsupervised rules. As opposed to the unsupervised rule, a supervised one could potentially facilitate the learning speed with the help of an instructor signal. Although so far there is no strong experimental confirmation of the supervisory signal, an increasing body of evidence shows that this kind of learning is also exploited by the brain [5].

Learning schemes focusing on processing spatiotemporal spikes in a supervised manner have been widely studied. With proper encoding methods, these schemes could be applied to image categorization. In [6], the spike-driven synaptic plasticity mechanism is used to learn patterns encoded by mean firing rates. A rate coding is used to encode images for categorization. The learning process is supervised and stochastic, in which a teacher signal steers the output neuron to a desired firing rate. According to this algorithm, synaptic weights are modified upon the arrival of pre-synaptic spikes, considering the state of post-synaptic neuron's potential and its recent firing activity. One of the major limitations of this algorithm is that it could not be used to learn patterns presented in the form of precise timing spikes. Different from the spike-driven synaptic plasticity, the tempotron learning rule [7] is efficient to learn spike patterns in which information is embedded in precise timing spikes as well as in mean firing rates. This learning rule modifies the synaptic weights such that a trained neuron fires once for patterns of corresponding category and keeps silent for patterns of other categories. The ReSuMe learning rule [8, 9] is also a supervised rule in which the trained neuron can fire at desired times when corresponding spatiotemporal patterns are presented. It has been demonstrated that the tempotron rule and the ReSuMe rule are equivalent under certain conditions [10].

Although SNNs show promising capabilities in achieving a performance similar to living brains due to their more faithful similarity to biological neural networks, one of the main challenges of dealing with SNNs is getting data into and out of them, which requires proper encoding and decoding methods. The temporal learning algorithms are based on spatiotemporal spike patterns. However, the problem remains how to represent real-world stimuli (like images) by spatiotemporal spikes for further computation in the spiking network. To deal with these problems, a unified systematic model, with consistent encoding, learning and readout, is required.

The main contribution of this chapter lies in the design of a unified system model of spiking neural network for solving pattern recognition problems. To the best of our knowledge, this is the first work in which complex classification task is solved through combination of biologically plausible encoding and supervised temporal learning. The system contains consistent encoding, learning and readout parts. Through the network, we fill the gap between real-world problem (image encoding) and theoretical studies of different learning algorithms for spatiotemporal patterns. Finally, our approach suggests a plausibility proof for a class of feedforward models of rapid and robust recognition in the brain.

2.2 The Spiking Neural Network

In this section, the feedforward computational model for pattern recognition is described. The model composes of 3 functional parts: the encoding part, the learning part and the readout part. Figure 2.1 shows the general architecture of the system model.

Considering the encoding, the latency code is a simple example of temporal coding. It encodes information in the timing of response relative to the encoding window, which is usually defined with respect to the stimulus onset. The external stimuli would trigger neurons to fire several spikes in different times. From biological observations, visual system can analyze a new complex scene in less than 150 ms [11, 12]. This period of time is impressive for information processing considering billions of

Fig. 2.1 A schematic of the feedforward computational model for pattern recognition. It contains three functional parts: encoding, learning and readout. A stimulus is converted into spatiotemporal spikes by the encoding neurons. This spiking pattern is passed to the next layer for learning. The final decision is represented by the readout layer

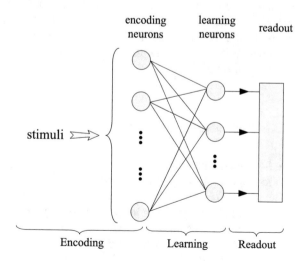

neurons involved. This suggests that neurons exchange only one or few spikes. In addition, it is shown that subsequent brain region may learn more and earlier about the stimuli from the time of first spike than from the firing rate [11].

Therefore, we use single spike code as the encoding mechanism. Within the encoding window, each input neuron fires only once. This code is simple and efficient, and the capability of encoding information in the timing of single spikes to compute and learn realistic data has been shown in [13]. Compared to rate coding as used in [6], this single spike coding would potentially facilitate computing speed since less spikes are involved in the computation.

Our single spike coding is similar to the rank order coding in [14, 15] but taking into consideration of the precise latency of the spikes. In the rank order coding, the rank order of neurons' activations is used to represent the information. This coding scheme is still under research. Taking the actual neurons' activations into consideration but not their rank orders, our proposed encoding method could convey more information than the rank order coding. Since this coding utilizes only a single spike to transmit information, it could also potentially be beneficial for efficient very large scale integration (VLSI) implementations.

In the learning layer, supervised rules are used since they could improve the learning speed with the help of the instructor signal. In this chapter, we investigate the tempotron rule and the ReSuMe rule.

The aim of the readout part is to extract information about the stimulus from the responses of learning neurons. As an example, we could use a binary sequence to represent a certain class of patterns in the case that each learning neuron can only discriminate two groups. Each learning neuron responds to a stimulus by firing (1) or not firing (0). Thus, the total N learning neurons as the output can represent a maximum number of 2^N classes of patterns.

A more suitable scheme for readout would be using population response. In this scheme, several groups are used and each group, containing several neurons, is one particular representation of the external stimuli. Different groups compete with each other by a voting scheme in which the group with the most amount of firing neurons would be the winner. This scheme is more compatible with the real brain since the information is presented by the cooperation of a group of neurons rather than one single neuron [16].

2.3 Single-Spike Temporal Coding

We have mentioned the function of the encoding layer is to convert stimulus into spatiotemporal spikes. In this section, we illustrate our encoding model of single-spike temporal coding, which is inspired from biological agents.

The retina is a particular interesting sensory area to study neural information processing, since its general structure and functional organization are remarkably well known. It is widely believed that information transmitted from retina to brain codes the intensity of the visual stimuli at every place in visual field. The ganglion cells (GCs) collect the information from their receptive fields which could best drive spiking responses [17]. In addition, different ganglion cells might have overlapped centers of receptive fields [18]. A simple encoding model of retina is described in [14] and is used in [15]. The GCs are used as the first layer in our model to collect information from original stimuli.

Focusing on emulating the processing in visual cortex, a realistic model (HMAX) for recognition has been proposed in [19] and widely studied [1, 20, 21]. It is a hierarchical system that closely follows the organization of visual cortex. The HMAX performs remarkably well with natural images by using alternate simple cells (S) and complex cells (C). Simple cells (S) gain their selectivity from a linear sum operation, while complex cells (C) gain invariance through a nonlinear max pooling operation. Like the HMAX model, in order to obtain an invariant encoding model to some extent, a complex cells (CCs) layer is used in our model. In the brain, equivalents of CCs may be in V1 and V4 (see [22] for more details).

In our model (see Fig. 2.2), the image information (intensity) is transmitted to GCs through photo-receptors. Each GC linearly integrates at its soma the information from its receptive field. Their receptive fields are overlapping and their scales are generally distributed non-uniformly over the visual field. DoG (difference of gaussian) filters are used in the GCs layer since this filter is believed to mimic how neural processing in the retina of the eye extracts details from external stimuli [23, 24]. Several different scales of DoG would construct different GCs images. The CCs unit would oper-

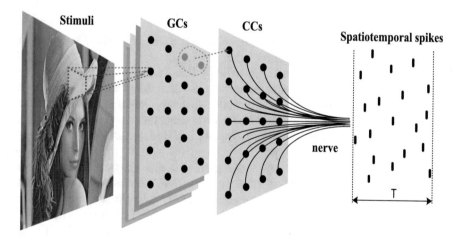

Fig. 2.2 Architecture of the visual encoding model. A gray-scale image (as the stimuli) is presented to the encoding layer. The photo-receptors transmit the image information analogically and linearly to the corresponding ganglion cells (GCs). Each ganglion cell collects information from its receptive field (an example shown as the *red dashed* box). There are several layers of GCs and each has a different scale of receptive field. The complex cells (CCs) collect information from a local position of GCs and a MAX operation among these GCs determines the activation value of CC unit. Each CC neuron would fire a spike according to their activations. These spikes are transmitted to the next layer as the spatiotemporal pattern in particular time window (T)

ate a nonlinear max pooling to obtain an amount of invariance. Max pooling over the two polarities, different scales and different local positions provides contrast reverse invariance, scale invariance and position invariance, respectively. Biophysically plausible implementations of the MAX operation have been proposed in [25], and biological evidences of neuron performing MAX-like behavior have been found in a subclass of complex cells in V1 [26] and cells in V4 [27].

The activation value of CC unit would trigger a firing spike. Strongly activated CCs will fire earlier, whereas weakly activated will fire later or not at all. The activation of the GC is computed through the dot product as:

$$GC_i := < I, \phi_i > = \sum_{l \in R_i} I(l) \cdot \phi_i(l) \tag{2.1}$$

where $I(l)$ is the luminance of pixel l which is sensed by the photo-receptor. R_i is the receptive field region of neuron i. ϕ_i is the weight of the filter.

The GCs compute the local contrast intensities at different spatial scales and for two different polarities: ON- and OFF-center filters. We use the simple DoG as our filter where the surround has three times the width of the center. The DoG has the form as:

$$DoG_{\{s,l_c\}}(l) = G_{\sigma(s)}(l - l_c) - G_{3 \cdot \sigma(s)}(l - l_c) \tag{2.2}$$

$$G_{\sigma(s)}(l) = \frac{1}{2\pi \cdot \sigma(s)^2} \cdot exp(-\frac{\|l\|^2}{2 \cdot \sigma(s)^2}) \tag{2.3}$$

where $G_{\sigma(s)}$ is the 2D Gaussian function with variance $\sigma(s)$ which depends on the scale s. l_c is the center position of the filter.

An example of the DoG filter is shown in Fig. 2.3. An OFF-center filter is simply an inverted version of an ON-center receptive field. All the filters are sum-normalized to zero and square-normalized to one so that when there is no contrast change in the image the neuron's activation would be zero and when the image is same with the filter the neuron's activation would be 1. Therefore, all the activations of the GCs are scaled to the same range ($[-1, 1]$).

The CCs max over different polarities according to their absolute values at same scale and same position. Through this max operation, the model gain a contrast reverse invariance (Fig. 2.4a). From the property of the polar filters, only one could be positive activated for a given image. Similarly, the scale invariance is increased by max pooling over different scales at the same position (Fig. 2.4b). Finally, the position invariance is increased by pooling over different local positions (Fig. 2.4c). The dimension of images is reduced since only the max activated value in a local position is preserved.

Figure 2.5 shows the basic processing procedures in different encoding layers. Through the encoding, the original image is sparsely presented in the CCs (Fig. 2.5f).

The final activations of CCs are used to produce spikes. Strongly activated neurons would fire earlier, whereas weakly activated ones would fire later or not at all. The

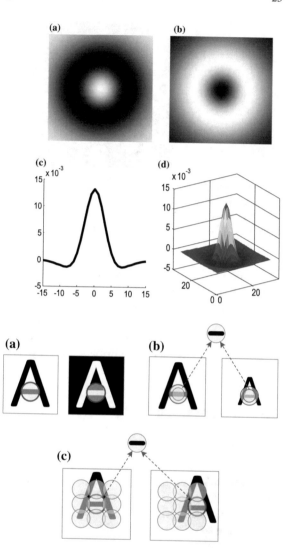

Fig. 2.3 Linear filters in retina. **a** is an image of the ON-center DoG filter, whereas (**b**) is an image of the OFF-Center filter. **c** is the one-dimensional show of the DoG weights and (**d**) is the 2-dimensional show

Fig. 2.4 Illustration of invariance gained from max pooling operation. **a** the contrast reverse invariance by max pooling over polarities. **b** the scale invariance by max pooling over different scales. **c** the local position invariance by max pooling over local positions. The *red circle* denotes the maximally activated one

spike latencies are then linearly mapped into a predefined encoding time window. These spatiotemporal spikes are transmitted to the next layer for computation.

In our encoding scheme, we consider the actual values of neurons' activations to generate spikes but not the rank order of these activations as used in [14, 15]. This could carry more information than the rank order coding which only considers the rank order of different activations and ignores the exact differences between different activations. For example, there are 3 neurons (n_1, n_2 and n_3) having their activations (C_1, C_2 and C_3) in the range of [0, 1]. Pattern P_1 is represented by ($C_1 = 0.1$, $C_2 = 0.3$ and $C_3 = 0.9$); pattern P_2 is represented by ($C_1 = 0.29$, $C_2 = 0.3$ and $C_3 = 0.32$). In rank order coding, it will treat P_1 and P_2 as same patterns since the

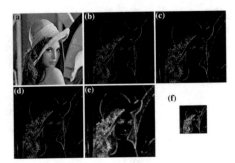

Fig. 2.5 Illustration of the processing results in different encoding procedures. **a** is the original external stimulus. **b** and **c** are the processing results in layer GCs with different scales. **d**, **e** and **f** are the processes in the CCs layer. **d** is the result of max pooling over different scales. **e** is max pooling over different local positions. **f** is the sub-sample from (**e**)

rank orders are same. For our encoding, in contrast, P_1 and P_2 would be treated as totally different patterns. In addition, the rank order coding would be very sensitive to the noise since the encoding time of one neuron depends on other neurons' rank. For example, if the least activated value is changed to a max activated value because of a disturbance, the rank of all the other neurons would be changed. However in our proposed algorithm only the information of the disturbed neuron would be affected.

2.4 Temporal Learning Rule

Temporal learning rule aims at dealing with information encoded by precise timing spikes. In this section, we consider supervised mechanisms like the tempotron rule and the ReSuMe rule that could be used for training neurons to discriminate between different spike patterns. Whether a LTP or LTD process occurs depends on the supervisory signal and the neuron's activity. This kind of supervisory signal can facilitate the learning speed compared to the unsupervised method.

2.4.1 The Tempotron Rule

In [7], the tempotron learning rule is proposed. According to this rule, the synaptic plasticity is governed by the temporal contiguity of a presynaptic spike and a postsynaptic depolarization, and a supervisory signal. The tempotron can make appropriate decisions under a supervisory signal by tuning fewer parameters.

In binary classification problem, each input pattern presented to the neuron belongs to one of two classes (which are labeled by P^+ and P^-). One neuron can make decision by firing or not. When a P^+ pattern is presented to the neuron, it should elicit a spike; when a P^- pattern is presented, it should keep silent by not firing. The tempotron rule modifies the synaptic weights (w_i) whenever there is an error. This rule performs like gradient-descent rule that minimizes a cost function as:

$$C = \begin{cases} V_{thr} - V_{t_{max}}, & \text{if the presented pattern is } P^+; \\ V_{t_{max}} - V_{thr}, & \text{if the presented pattern is } P^-. \end{cases} \qquad (2.4)$$

where $V_{t_{max}}$ is the maximal value of the post-synaptic potential V.

Applying the gradient descent method to minimize the cost leads to the tempotron learning rule:

$$\Delta w_i = \begin{cases} \lambda \sum_{t_i < t_{max}} K(t_{max} - t_i), & \text{if } P^+ \text{ error}; \\ -\lambda \sum_{t_i < t_{max}} K(t_{max} - t_i), & \text{if } P^- \text{error}; \\ 0, & \text{otherwise}. \end{cases} \qquad (2.5)$$

where t_{max} denotes the time at which the neuron reaches its maximum potential value in the time domain. $\lambda > 0$ is a constant representing the learning rate. It denotes the maximum change on the synaptic efficacies. P^+ error denotes that the neuron should fire but it did not; P^- error denotes that the neuron should not fire but it did. According to the kernel shape, the efficacies of afferents that spike near to t_{max} change more than that are away from it. In this rule, t_{max} is the reference time for updating synaptic weights.

2.4.2 The ReSuMe Rule

The ReSuMe described in [9] is a supervised method that aims to produce desired spike trains in response to the given input sequence. According to this rule, the synaptic weights are modified according to the following equation:

$$\frac{d\omega_i(t)}{dt} = \lambda[S^d(t) - S^{out}(t)][a + \int_0^\infty W(s)S^i(t-s)ds] \qquad (2.6)$$

where λ is the learning rate, a is a constant, W is a learning window with a exponential form $(W(s) = Ae^{-s/\tau_E})$. $S^d(t)$, $S^{out}(t)$ and $S^i(t)$ are the target, post- and pre-synaptic spike trains, respectively. Although the shape of learning window is not restricted to exponential form, this shape can result in a better performance of convergence [28]. The spike trains have the following form:

$$S(t) = \sum_{f=1}^n \delta(t - t^f) \qquad (2.7)$$

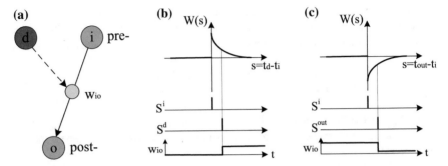

Fig. 2.6 Illustration of the ReSuMe learning rule. **a** demonstrates that the synaptic plasticity depends on the correlation between the pre- and postsynaptic firing times, and on the correlation between pre- and desired firing times. **b** demonstrates that the synaptic weight is potentiated whenever a desired spike is observed. **c** shows that the synaptic weight is depressed whenever the trained neuron fires. This figure is revised from [8]

where t^f denotes the moment of the f-th spike in the train, n denotes the total number of spikes in the train, $\delta(x)$ is the impulse function $\delta(x) = 1$ if $x = 0$ (or 0 otherwise).

Figure 2.6 illustrates the ReSuMe learning rule. The synaptic efficacy depends not only on the correlation between the pre-synaptic and post-synaptic firing times but also on the correlation between the pre-synaptic and desired firing times. A desired spike would result in synaptic potentiation, and a post-synaptic spike would result in synaptic depression.

After a learning trial, the total synaptic change is:

$$\Delta\omega_i = \lambda a(n^d - n^{out}) + \lambda \Sigma_{t^d} \Sigma_{t_i \leq t^d} W(t^d - t_i) \qquad (2.8)$$
$$- \lambda \Sigma_{t^{out}} \Sigma_{t_i \leq t^{out}} W(t^{out} - t_i)$$

where n^d and n^{out} are the number of spikes from the desired and the actual output spike trains respectively. t_i is the pre-synaptic spike time.

The ReSuMe rule could be used for both the batch learning and the online learning.

2.4.3   The Tempotron-Like ReSuMe Rule

As proposed in [10], the tempotron learning rule is a particular case of ReSuMe rule under certain conditions. The rule discussed here is a connection between the tempotron rule and the ReSuMe rule.

Considering to apply ReSuMe to the tempotron setup, the combined rule can be approached. The neuron is only allowed to fire once or not. After a spike is emitted, the neuron shunts all its incoming spikes immediately. If there is only one spike, regardless of its time, it is reasonable to consider the neuron firing at t_{max}. This learning rule follows [10]:

$$\Delta w_i = \begin{cases} \lambda a + \lambda \sum_{t_i \leq t_{max}} W(t_{max} - t_i), & \text{if } n^d = 1, n^{out} = 0; \\ -\lambda a - \lambda \sum_{t_i < t_{out}} W(t_{out} - t_i), & \text{if } n^d = 0, n^{out} = 1; \\ 0, & \text{if } n^d = n^{out}. \end{cases} \quad (2.9)$$

When $a = 0$ and $W(s) = K(s)$, the combined rule is equivalent to the tempotron learning rule. This implicates that the tempotron rule is a particular case of the ReSuMe rule.

2.5 Simulation Results

In this section, several simulations are performed to test the performance of the network and different learning rules.

2.5.1 The Data Set and the Classification Problem

The stimuli from real world typically have a complex statistical structure. It is quite different from idealized case of random patterns often considered. In the real world, the stimuli hold large variability in a given class and have a high level of correlation between members of different classes. The data set we considered here is the MNIST digits (see Fig. 2.7).

The MNIST data set contains a large number of examples of hand-written digits, which consists of ten classes (digits 0 to 9) of examples and each example is an image of 28×28 pixels. The MNIST data set is available from http://yann.lecun. com/exdb/mnist, where many classification results from different methods are also listed. All images from this data set are gray-scale.

Fig. 2.7 Examples of handwritten digits from MNIST dataset

2.5.2 Encoding Images

Each image is presented to the encoding layer, and is then converted into spatiotemporal pattern. We use the coding strategy discussed previously through which the output is sparse, as is observed in biological agents [29].

For simplicity of applying the encoding algorithm to the data set, we distribute GCs with different receptive fields all over the image (each pixel). The image size in GCs is same as the input image. Considering examples of 28-by-28 images, we choose two scales for the filters ($\sigma = 1$ for 5×5 pixels as scale 1, and $\sigma = 2$ for 7×7 pixels as scale 2). The CCs layer performs the max pooling operation on the previous GCs layer. For local position operation we choose 6×6 pixels and we set the overlap pixels to be 3 in one axis (x or y) for sub-sampling operation. A detailed process of max operation is described in [19].

The application of all these processes produces a set of analog values, corresponding to the activation levels of our CCs unit. The strongly activated cell will fire earlier, whereas the weakly activated will fire later or not at all. The spike latencies are linearly mapped into a predefined encoding time window (100 ms in this study). The activation values are linearly converted to delay times, associating $t = 0$ with activation value 1 and later times up to 100 ms with lower activation values. The neurons with activation value of 0 (or below a chosen small value) will not fire due to the weak activation.

An illustration of encoding an image is shown in Fig. 2.5. Our scheme is to extract the basic information and encode it to a spatiotemporal spike pattern. Through the whole encoding structure, a sparse representation of the original incoming image is finally obtained. Using this sparse representation to generate the spike pattern would, to some extent, be compatible with biological observations in retina.

2.5.3 Choosing Among Temporal Learning Rules

In the tempotron rule, we specify the following parameters. The ratio between the membrane and the synaptic constants is fixed at $\tau_m / \tau_s = 4$. The threshold V_{thr} is set to 1 and V_{rest} is set to 0. We use $\tau_m = 10$ ms and $\lambda = 0.002$.

For comparison purpose, in the ReSuMe we use the similar neuron model as the one in the tempotron rule. However, the difference is that when the neuron emits a spike, its potential is reset to a rest value (0 here) and is hold there for a refractory period (3 ms here).

Since the ReSuMe rule is based only on the spiking times, it could work independently on the used spiking neuron models [9]. To verify the suitability of this rule for our chosen neuron model, we generate a spike pattern and force the neuron to respond at desired times. We choose 300 afferent synapses and each fires only once in the time window. The timing of each spike is generated randomly with uniform

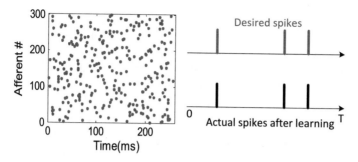

Fig. 2.8 Illustration of the suitability of ReSuMe rule for the chosen neuron model. The input pattern contains 300 afferent synapses and each fires once only. These spikes are generated randomly with uniform distribution. For the desired spike, 3 random spiking times are chosen

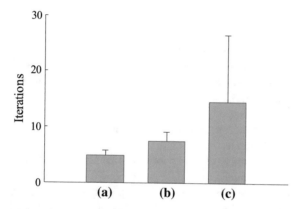

Fig. 2.9 The number of iterations needed for the correct classification of spike patterns, through different learning rules. **a** is the tempotron learning rule. **b** is the tempotron-like ReSuMe rule. **c** is the ReSuMe rule in which if the neuron fires, it should spike at a desired time. Over 100 experiments with different initial conditions, the averages (4.95, 7.36 and 14.48) and standard deviations (0.8454, 1.7438 and 12.014) are obtained for (**a**), (**b**) and (**c**), respectively

distribution between 0 and T. After learning, the neuron could perform as the desired way (see Fig. 2.8).

To compare the learning speed of different learning rules, we generate 30 spatiotemporal patterns and each pattern contains 120 afferent synapses. The spiking times are generated randomly with a uniform distribution between 0 and T. We randomly choose 3 patterns as one category that is needed to be discriminated from others. We record the minimum times of iterations for different rules to learn these patterns correctly. We perform this experiment for 100 times and the results are shown in Fig. 2.9.

According to Fig. 2.9, there is no significant difference of learning speed between the tempotron rule and tempotron-like ReSuMe rule. This is because the only difference between these two rules is the kernel windows which have a similar effect on the synaptic change. However, compared to the ReSuMe rule, the tempotron rule is much faster (about 3 times as the ReSuMe rule). Besides this, the learning speed of the ReSuMe varies significantly for different initial conditions (such as the number of patterns, the initial weights and the learning rate). For the sake of fast recognition, we choose the tempotron rule as our learning rule.

2.5.4   The Properties of Tempotron Rule

Since the tempotron rule is chosen, a test on its properties is needed.

2.5.4.1   Capacity

As is used for perceptron [30], the ratio of the number of random patterns (N_p) that correctly classified by the neuron over the number of its synapses (N_{in}), $\alpha = N_p/N_{in}$, is used to measure the load of the neuron. An important characteristic of neuron's capacity is the maximum load that it can learn. As studied in [7], the maximum recognition load of a tempotron can reach 3 approximately, which means that the number of patterns the neuron can learn could roughly approach to 3 times the number of synapses connected to it.

For our chosen neuron, a test on its load is shown in Fig. 2.10. We set $N_{in} = 100$ and generate different number of spike patterns within a fixed time window ($T = 100$ ms). Each afferent fires only once and the spiking time is randomly chosen from uniform distribution within T. The mean number of cycles of pattern presentations for error-free classification is shown versus the load (α). Although a more robust estimation of the load is feasible by allowing a small percentage of false alarms, the

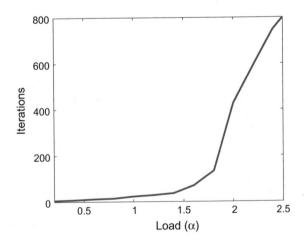

Fig. 2.10 The mean number of iterations of pattern presentations for error-free classification versus neuron load. The patterns are randomly generated within the fixed time window (100 ms). The number of synapses is 100. Data are averaged over 20 runs

rigorous condition of error-free classification is useful to testify the neuron's ability of classifying all assigned patterns successfully.

According to Fig. 2.10, the neuron could successfully learn the patterns within several tens of iterations if the load is not very high (below 1.5), but the number of iterations would increase sharply when the load is over 1.5. This means that under a higher load the neuron needs more time to learn the patterns or the learning process could never converge.

This load test, to some extent, could guarantee the learning convergence when the tempotron neuron is applied to our chosen recognition task. In our task, there are only ten categories and patterns in each category share some common features. Compared to the randomly generated patterns, the neuron's capacity might be sufficient to learn these real-world stimuli.

2.5.4.2 Robustness

In some cases, the external noise might change the encoded spike patterns more or less. The tempotron rule should hold some level of robustness to tolerate the noise. To assess the robustness of the learning rule, we trained the neuron with a number of patterns ($\alpha = 1$). Then we tested the performance of the neuron when facing with jittered versions of previous learned patterns. The jittered pattern was generated by adding a Gaussian noise to all spike times of a template pattern. The robust performance of the neuron is shown in Fig. 2.11.

According to Fig. 2.11, the performance of correct recognition decreases with increasing jitter. Within a limited jitter range (0–3 ms), the performance stays in a relatively high level (over 0.8). This indicates the learning rule is robust to the presence of temporal noise to some extent.

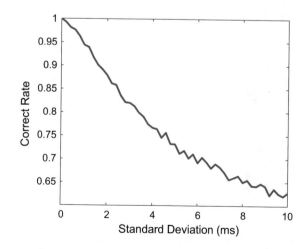

Fig. 2.11 The mean correct rate of classification on jittered spike patterns. The jittered pattern is generated by adding Gaussian noise with standard deviation to all spike times of a template pattern

2.5.5 Recognition Performance

The combined system is applied to recognize different patterns. To see the ability of our system network on the recognition task, we use a small data set from the MNIST (50 digits and 5 for each category). And we choose four neurons as the readout. We call this readout as the fully distributed scheme with no redundancy (each neuron codes for one bit). After several iterations of training, the network can recognize all the patterns in this data set. Here, we take the recognition results of several digits as an example (Fig. 2.12). If the potential of the learning neuron crosses the threshold, namely it fires, the value of this neuron is considered as 1, otherwise it is 0. In Fig. 2.12, when image "0" shows up to the network, none of the learning neurons fire, so the result is [0000]. For image "3", the result is [0011], and for "9" it's [1001]. This indicates that the tempotron rule applied in our model could recognize different classes of images successfully.

However, using only four neurons as the readout in a binary format might be very sensitive to changes of input images, especially considering the real-world stimuli in which samples hold large variability in a given class and overlap with members in different classes. If only one neuron misclassified the incoming pattern while others correctly responded, the pattern was still wrongly classified. Researchers have found that neighboring neurons have similar response properties. Depending on this, neural groups are used for assembly computing [16].

Thus, we use several grouped pools as our readout. We firstly consider a distributed code with redundancy: 4 pools of 20 neurons each. Each pool codes for one binary

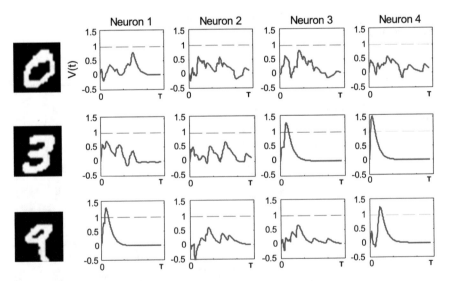

Fig. 2.12 Recognition results of digits by Tempotron learning rule. Here shows 4 learning neurons (Neuron 1–4) and 3 images. The neuron responds to an image by firing (1) or not (0). The results for "0", "3" and "9" are [0000], [0011] and [1001], respectively

feature as in Fig. 2.12. A voting system decides if the binary feature is 0 or 1 based on the voting majority in this pool. Then we consider a localist scheme with redundancy, where each pool of 20 neurons codes for only one category. For an incoming stimulus, it is classified into a category according to the pool which has the most amount of voting neurons fired. If two or more pools have the same maximal firing number, the incoming stimulus is classified as unknown pattern.

These two schemes of readout with redundancy are used. For cross-validation, we choose 500 digits (50 images for each category) as our training set and randomly choose other 100 images from the MNIST data set as the testing set. In the training phase, each neuron is trained with a sub-training set chosen from the training set. This sub-training set consists of examples randomly chosen from the corresponding category and also other categories. After training, the performance is tested on both training set and testing set. The correct rate on the testing set is around 50% for the distributed code with no redundancy, and is around 79% for the localist code with redundancy. For distributed code, although the robustness for coding one bit feature is improved comparing to single neuron code, it is still not comparable to the localist one. This is due to that in the distributed code the final decision highly depends on correct reaction of each pool, but in the localist code it only depends on a correct major voting of one corresponding pool. Thus, in the localist scheme, the robustness is not only due to the redundancy but also to the localist aspect. This localist scheme is considered in our following experiments.

To make a comparison with the benchmark machine learning method, SVMs are chosen to perform the classification on the CCs activation values. Since SVM also has a binary decision behavior, we set the same classification condition on training and testing as for tempotron. The performances of both the tempotron and SVM on the training set and testing set are shown in Fig. 2.13. The corresponding recognition rates are shown in Table 2.1.

According to Fig. 2.13, our network with the tempotron rule performs at a high correct rate (around 93.7%) on the training set and at an acceptable correct rate (around 79%) on the testing set, especially considering the small data set (500 images) used for training. Comparing with SVM under the same condition of our encoding model, the performances of spiking neurons are better than SVM for the training set and comparable to SVM for the testing set. From a biological point of view, our system attempts to perform robust and rapid recognition with a brain-like architecture.

Table 2.1 The classification performance of tempotron and SVM on MNIST

	Tempotron rule		SVM	
Percentage(%)	Training	Testing	Training	Testing
Correct rate	93.67 ± 0.67	78.5 ± 1.85	90.24 ± 0.98	79.33 ± 2.03
Wrong rate	4.48 ± 0.58	18.35 ± 1.85	6.88 ± 0.78	18.15 ± 1.69
Unknown rate	1.86 ± 0.61	3.15 ± 1.64	2.89 ± 0.86	2.53 ± 2.04

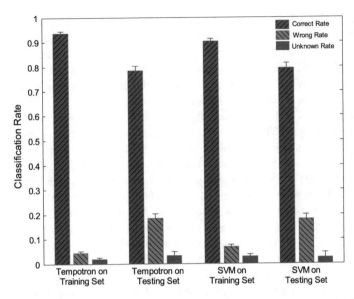

Fig. 2.13 The classification performance of tempotron and SVM. The system is trained 40 times each for tempotron and SVM. After each training time, the generalization is performed on both the training and testing set. The averages and standard deviations are plotted

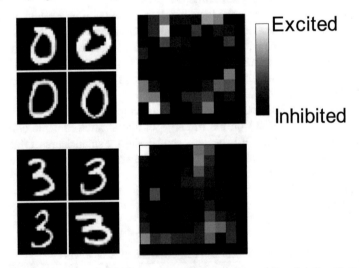

Fig. 2.14 Average weights of the spiking neurons in the pool representing digit 0 and 3. *Left* Image samples of digit 0 and 3 from MNIST are listed. *Right* The picture of the average weight of the spiking neurons in corresponding group. Inhibited afferents are plotted *black*, while excited ones are plotted *gray*-scale according to their weights

To investigate the states of the spiking neurons in one grouped pool after learning, a picture of the average weights is shown in Fig. 2.14. According to Fig. 2.14, the grouped neurons, cooperating together, roughly grab a general and basic feature of the learned category. Taking digit 0 as an example, the center weights are mostly inhibited since these neurons are rarely activated by the incoming stimulus 0 through our encoding model.

2.6 Discussion

Discussions on the proposed system are given as follows.

2.6.1 Encoding Benefits from Biology

Through the layers of GCs and CCs the external stimuli are sparsely represented in the activation values of CCs units. These activation values are used to generate spiking patterns in a time domain. It already has been shown that coding schemes based on the firing rates are unlikely to be efficient enough for fast information processing [14, 31]. Considering the rapid processing in the brain and billions of neurons involved, a temporal code which uses single spikes is, in principle, capable of carrying substantial information about the external stimuli [11] and facilitating the computational speed. In several sensory systems, shorter latencies of spikes result from stronger stimulation [32, 33]. In our encoding layer, the strongly activated neurons would fire earlier, whereas the weakly activated neurons would fire later or not at all. The chosen encoding window of the temporal patterns is on a scale of hundreds of milliseconds, which matches the biologically experimental results as mentioned in [34–36]. In addition, our encoding is efficient and the spiking output is sparse as observed in biological retinas [29, 37].

2.6.2 Types of Synapses

The types of synapses are determined by the signs of their efficacies, with positive values corresponding to excitatory synapses and negative values to inhibitory synapses. Although this model is far from biological realism, it is proved to be a useful computational approach [9]. In the neuron model, the sign of synapse could change by learning. The learning also works when the signs of synapses are not allowed to change, but the capacity is reduced. For a practical usage for multiple-class problem, changing sign is allowed in the neuron model. This can be realized by altering the balance between excitatory and inhibitory pathways [7].

2.6.3 Schemes of Readout

Using a binary version of readout, the network is shown to be capable to finish a simple recognition task on a small data set. However, this kind of readout would be very sensitive to each neuron's performance in the readout. If only one neuron misclassifies the pattern while others do a correct classification, the final readout would also be wrong since it depends on all the neurons in a binary form. Using grouped pools could effectively compensate this. In nervous systems such as visual cortical areas [38] and hippocampus [39], information is commonly expressed through populations or clusters of cells rather than through single cell [40]. This strategy is robust since damage to a single cell will not have a catastrophic effect on the whole population. Through learning, neurons in the same group try to find the common features discriminating that category, and through voting, the most active group would be chosen. Another meaningful aspect of the readout is that there is an unknown decision. Since some samples in one category are quite similar to other categories (for example the digit "5" in the second row of Fig. 2.7), it is reasonable to label them as unknown rather than wrong. A further processing could be done for these unknown samples.

2.6.4 Extension of the Network for Robust Sound Recognition

In addition to the recognition on images, we also proposed a SNN for recognizing sounds. The general structure remains the same, where functional parts of encoding, learning and readout are involved. The major difference of the two systems is the encoding part. With a proper encoding scheme for sounds, the SNN can perform the

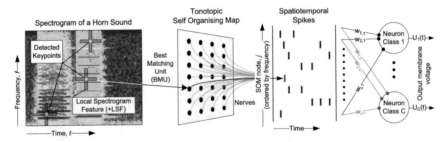

Fig. 2.15 The proposed LSF-SNN system for sound recognition. Firstly the keypoints are detected and the corresponding LSFs are extracted. Then, the SOM map is used to produce the output spatiotemporal spike patterns. These patterns are then learnt by the tempotrons for recognition

recognition well. We propose a novel approach based on the temporal coding of Local Spectrogram Features (LSF) [41], which generates spikes that are used to train the following neurons. The general structure for sound recognition is shown in Fig. 2.15. Our experiments demonstrate the robust performance of this system across a variety of noise conditions, such that it is able to outperform the conventional frame-based baseline methods. More details can be found in [41].

2.7 Conclusion

A systematic computational model by using consistent temporal encoding, learning and readout has been presented to explore brain-based computation especially in the regime of pattern recognition. It is a preliminary attempt to perform rapid and robust pattern recognition from a biological point of view. The schemes used in this model are efficient and biologically plausible. The external stimuli are sparsely represented after our encoding and the representations have properties of selectivity and invariance. Through the network, the temporal learning rules can be applied to processing real-world stimuli.

References

1. Serre, T., Oliva, A., Poggio, T.: A feedforward architecture accounts for rapid categorization. Proc. Natl. Acad. Sci. **104**(15), 6424–6429 (2007)
2. Perrett, D.I., Hietanen, J.K., Oram, M.W., Benson, P.J.: Organization and functions of cells responsive to faces in the temporal cortex. Philos. Trans. R. Soc. Lond. Ser. B **335**, 23–30 (1992)
3. Hung, C.P., Kreiman, G., Poggio, T., DiCarlo, J.J.: Fast readout of object identity from macaque inferior temporal cortex. Science **310**(5749), 863–866 (2005)
4. Tsukada, M., Pan, X.: The spatiotemporal learning rule and its efficiency in separating spatiotemporal patterns. Biol. Cybern. **92**, 139–146 (2005)
5. Knudsen, E.I.: Supervised learning in the brain. J. Neurosci. **14**(7), 3985–3997 (1994)
6. Brader, J.M., Senn, W., Fusi, S.: Learning real-world stimuli in a neural network with spike-driven synaptic dynamics. Neural Comput. **19**(11), 2881–2912 (2007)
7. Gütig, R., Sompolinsky, H.: The tempotron: a neuron that learns spike timing-based decisions. Nature Neurosci. **9**(3), 420–428 (2006)
8. Ponulak, F.: ReSuMe-new supervised learning method for spiking neural networks. Institute of Control and Information Engineering, Poznoń University of Technology, Technical Report (2005)
9. Ponulak, F., Kasinski, A.: Supervised learning in spiking neural networks with resume: sequence learning, classification, and spike shifting. Neural Comput. **22**(2), 467–510 (2010)
10. Florian, R.V.: Tempotron-Like Learning with ReSuMe. In: Proceedings of the 18th International Conference on Artificial Neural Networks. Part II, ICANN'08, pp. 368–375. Springer, Heidelberg (2008)
11. Gollisch, T., Meister, M.: Rapid neural coding in the retina with relative spike latencies. Science **319**(5866), 1108–1111 (2008)
12. Thorpe, S., Fize, D., Marlot, C.: Speed of processing in the human visual system. Nature **381**(6582), 520–522 (1996)

13. Bohte, S.M., Bohte, E.M., Poutr, H.L., Kok, J.N.: Unsupervised clustering with spiking neurons by sparse temporal coding and multi-layer RBF networks. IEEE Trans. Neural Netw. **13**, 426–435 (2002)
14. Van Rullen, R., Thorpe, S.J.: Rate coding versus temporal order coding: what the retinal ganglion cells tell the visual cortex. Neural Comput. **13**(6), 1255–1283 (2001)
15. Perrinet, L., Samuelides, M., Thorpe, S.J.: Coding static natural images using spiking event times: do neurons cooperate? IEEE Trans. Neural Netw. **15**(5), 1164–1175 (2004)
16. Ranhel, J.: Neural assembly computing. IEEE Trans. Neural Netw. Learn. Syst. **23**(6), 916–927 (2012)
17. Hubel, D.H., Wiesel, T.N.: Receptive fields and functional architecture of monkey striate cortex. J. physiol **195**(1), 215–243 (1968)
18. Burkart, Fischer: Overlap of receptive field centers and representation of the visual field in the cat's optic tract. Vis. Res. **13**(11), 2113–2120 (1973)
19. Riesenhuber, M., Poggio, T.: Hierarchical models of object recognition in cortex. Nature Neurosci. **2**(11), 1019–1025 (1999)
20. Masquelier, T., Thorpe, S.J.: Unsupervised learning of visual features through spike timing dependent plasticity. PLoS Comput. Biol. **3**(2) (2007)
21. Serre, T., Wolf, L., Bileschi, S., Riesenhuber, M., Poggio, T.: Robust object recognition with cortex-like mechanisms. IEEE Trans. Pattern Anal. Mach. Intell. **29**, 411–426 (2007)
22. Serre, T., Kouh, M., Cadieu, C., Knoblich, U., Kreiman, G., Poggio, T.: A theory of object recognition: computations and circuits in the feedforward path of the ventral stream in primate visual cortex. In: AI Memo (2005)
23. Enroth-Cugell, C., Robson, J.G.: The contrast sensitivity of retinal ganglion cells of the cat. J. Physiol. **187**(3), 517–552 (1966)
24. McMahon, M.J., Packer, O.S., Dacey, D.M.: The classical receptive field surround of primate parasol ganglion cells is mediated primarily by a non-GABAergic pathway. J. Neurosci. **24**(15), 3736–3745 (2004)
25. Yu, A.J., Giese, M.A., Poggio, T.: Biophysiologically plausible implementations of the maximum operation. Neural Comput. **14**(12), 2857–2881 (2002)
26. Lampl, I., Ferster, D., Poggio, T., Riesenhuber, M.: Intracellular measurements of spatial integration and the MAX operation in complex cells of the cat primary visual cortex. J. Neurophysiol. **92**(5), 2704–2713 (2004)
27. Gawne, T.J., Martin, J.M.: Responses of primate visual cortical neurons to stimuli presented by flash, saccade, blink, and external darkening. J. Neurophysiol. **88**(5), 2178–2186 (2002)
28. Ponulak, F.: Analysis of the resume learning process for spiking neural networks. Appl. Math. Comput. Sci. **18**(2), 117–127 (2008)
29. Olshausen, B.A., Field, D.J.: Sparse coding with an overcomplete basis set: a strategy employed by V1? Vis. Res. **37**(23), 3311–3325 (1997)
30. Gardner, E.: The space of interactions in neural networks models. J. Phys. **A21**, 257–270 (1988)
31. Gautrais, J., Thorpe, S.: Rate coding versus temporal order coding: a theoretical approach. Biosystems **48**(1–3), 57–65 (1998)
32. Reich, D.S., Mechler, F., Victor, J.D.: Independent and redundant information in nearby cortical neurons. Science **294**, 2566–2568 (2001)
33. Greschner, M., Thiel, A., Kretzberg, J., Ammermüller, J.: Complex spike-event pattern of transient ON-OFF retinal ganglion cells. J. Neurophysiol. **96**(6), 2845–2856 (2006)
34. Panzeri, S., Brunel, N., Logothetis, N.K., Kayser, C.: Sensory neural codes using multiplexed temporal scales. Trends Neurosci. **33**(3), 111–120 (2010)
35. Butts, D.A., Weng, C., Jin, J., Yeh, C.I., Lesica, N.A., Alonso, J.M., Stanley, G.B.: Temporal precision in the neural code and the timescales of natural vision. Nature **449**(7158), 92–95 (2007)
36. Borst, A., Theunissen, F.E.: Information theory and neural coding. Nature Neurosci. **2**(11), 947–957 (1999)
37. Hunt, J.J., Ibbotson, M.R., Goodhill, G.J.: Sparse coding on the spot: spontaneous retinal waves suffice for orientation selectivity. Neural Comput. **24**(9), 2422–2433 (2012)

38. Usrey, W., Reid, R.: Synchronous activity in the visual system. Annu. Rev. Physiol. **61**(1), 435–456 (1999)
39. Wilson, M., McNaughton, B.: Dynamics of the hippocampal ensemble code for space. Science **261**(5124), 1055–1058 (1993)
40. Pouget, A., Dayan, P., Zemel, R.: Information processing with population codes. Nature Rev. Neurosci. **1**(2), 125–132 (2000)
41. Dennis, J., Yu, Q., Tang, H., Tran, H.D., Li, H.: Temporal coding of local spectrogram features for robust sound recognition. In: 2013 IEEE International Conference on Acoustics, Speech and Signal Processing (ICASSP), pp. 803–807 (2013)

Chapter 3
A Spike-Timing Based Integrated Model for Pattern Recognition

Abstract During the last few decades, remarkable progress has been made in solving pattern recognition problems using network of spiking neurons. However, the issue of pattern recognition involving computational process from sensory encoding to synaptic learning remains underexplored, as most existing models or algorithms only target part of the computational process. Furthermore, many learning algorithms proposed in literature neglect or pay little attention to sensory information encoding, which makes them incompatible with neural-realistic sensory signals encoded from real-world stimuli. By treating sensory coding and learning as a systematic process, we attempt to build an integrated model based on spiking neural networks (SNNs), which performs sensory neural encoding and supervised learning with precisely timed sequences of spikes. With emerging evidence of precise spike-timing neural activities, the view that information is represented by explicit firing times of action potentials rather than mean firing rates has received increasing attention recently. The external sensory stimulation is first converted into spatiotemporal patterns using latency-phase encoding method and subsequently transmitted to the consecutive network for learning. Spiking neurons are trained to reproduce target signals encoded with precisely timed spikes. It is shown that using a supervised spike-timing based learning, different spatiotemporal patterns are recognized by different spike patterns with a high time precision in milliseconds.

3.1 Introduction

Everyday we recognize plenty of familiar and novel objects. However, we know little about the underlying mechanism of the sophisticated computation involved in the recognition process of human nervous system. Throughout our brain, neurons propagate information by generating clusters of electrical impulses called action potentials (APs) [1]. Analogue stimuli are encoded into spatiotemporal patterns and the neural representation of external world is the basis for perception and reaction [2]. Different encoding methods have been proposed by researchers, and among these approaches rate-based encoding (rate codes) and spike-based encoding (temporal codes) are the most widely studied coding schemes [3, 4]. Traditionally, it is believed that

© Springer International Publishing AG 2017
Q. Yu et al., *Neuromorphic Cognitive Systems*, Intelligent Systems
Reference Library 126, DOI 10.1007/978-3-319-55310-8_3

information is carried by the temporal average of spikes [5–7], and rate-based coding has been widely used in previous learning models such as performing stochastic gradient learning [8] and solving recognition problem relying on variance of input currents [9]. Although rate codes work well when the stimulus is constant or varying slowly, which is not common in real-world stimulations. Unlike the rate coding, temporal encoding schemes assume that information is carried by the precisely timed spikes, which provides more information capacity than the mean firing rate of neurons [10, 11]. It has been found that temporally varying sensory information such as visual and auditory signals is processed and stored with high precision in brain [12, 13], and precisely timed spikes are important for the integration process of cortical neurons [14]. Therefore, temporal codes can describe neural signal more precisely which enable us to exploit time as a resource for communication and computation in spiking neural networks.

Recent neurophysiological results show that the precision of temporal spikes may be triggered by the rapid intensity transients [15] and even a single spike can carry substantial information about visual stimuli [16]. The low response variability of retinal ganglion cells shows that the most important information of a firing event generated by visual neurons may be reserved by the time of the first spike and the number of spikes [17]. Furthermore, experimental results show that most information carried by spikes is the timing of the first spike after stimulus onset [16]. In human retina, visual signal from 10^8 photoreceptor cells are projected to 10^6 retinal ganglion cells (RGCs) in the form of spike trains [15]. Hence the information compression is indispensable during the projection. In addition, action potentials have been shown to be related to the phases of the intrinsic sub-threshold membrane potential oscillations [18, 19]. The phase locking between action potential and gamma oscillation has also been discovered in electric fish [20] and the entorhinal cortex [21]. Phase coding has been successfully utilized to perform sequences learning and episodic memory in hippocampus via phase precession in previous works [22–24]. The phase information of spikes is exploited within each receptive field. As each ganglion cell receives information from the photoreceptor cells in its receptive field, phase coding is used to reserve spatial information during compression as described in Sect. 2.2. Thus, we believe that the combination of temporal and phase coding offers a new way to implement the compression as well as to explain the compression process.

After sensory encoding, the neural system needs to learn neural signals that represent external sensory stimulation. Spike-based learning algorithms compute with firing times and make use of the inter spike intervals so that they are compatible with temporal codes. Hebbian synaptic plasticity has been viewed as the basic mechanism for learning and memory [25, 26], in which the synaptic efficacy is increased if the presynaptic neuron repeatedly contributes to the firing of postsynaptic neuron. As precise spike timing [27] and relative timing between pre- and post-synaptic firing [28] are discovered, learning with millisecond precision has received intensive interests. Spike-timing-dependent plasticity (STDP) is believed to play an important role in learning, memory and the development of neural circuits [29]. However, many existing learning models use rate codes as the neural representation of information, and learning with temporal codes remains an open research topic. The objective of learning is to train output neurons to respond selectively to inputs and generate desired output spike patterns by adjusting synaptic plasticity. Since the membrane potential of postsynaptic neuron is determined by the spikes of afferent neurons, the generation of postsynaptic spike is the result of the cooperative integration and synchronization of presynaptic input spikes [30, 31]. When the input spikes arrive in synchrony and a sufficiently large depolarization of postsynaptic membrane potential is achieved, a firing event will be triggered. Since we consider explicit desired patterns for recognition task, supervised learning is preferred due to its efficiency and accuracy. Moreover, growing evidences indicate that supervised learning is also employed in cerebellum and cerebellar cortex [32, 33]. It has also been demonstrated to be a successful form of learning to establish network with cognition functions [34, 35]. We adopt a spike-timing based supervised learning algorithm recently developed by [36], in which the error between the target spike train and the actural one is used as the supervisory signal. In addition, the firing intervals between pre- and postsynaptic spikes are recorded for synaptic plasticity modification, through which the actual output patterns approximate the desired output patterns gradually.

The contribution of this work is to bridge the gap between sensory encoding and synaptic information processing by proposing an integrated computational model with spike-timing based encoding scheme and learning algorithm. This helps to reveal the neural mechanisms starting from visual encoding to synaptic learning and the computational process in central nervous system. Such an encoding and learning algorithms in the proposed spike-based model are integrated in a consistent scheme: temporal codes. The encoding method provides a possible mechanism for converting visual information into neural signals. The spiking neurons are trained to classify spatiotemporal patterns based on the temporal configuration of spikes rather than firing rates of neurons.

This chapter is organized as follows: In Sect. 3.2, we introduce the general structure, encoding method and learning algorithm of the proposed integrated model. In Sect. 3.3, the performance and properties of the integrated model are demonstrated by numerical simulations. Section 3.4 reviews the related works while Sect. 3.5 concludes and discusses the limitations and extensions of the integrated model proposed in this work.

3.2 The Integrated Model

3.2.1 Neuron Model and General Structure

In our proposed integrated model, all neurons are modeled with the leaky integrate-and-fire (LIF) model [37], which is defined as:

$$\tau \frac{dV}{dt} = -(V - V_r) + R(I_0 + I_{in} + I_n) \tag{3.1}$$

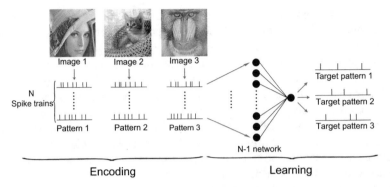

Fig. 3.1 General structure and information process of the integrated model. The main components of the model are the encoding part and the learning part. The spike-based model employs temporal codes as the neural representation of external information. The latency-phase encoding as discussed in Sect. 2.2 is used to convert the image into spatiotemporal patterns consisting of N spike trains. After sensory encoding, each spike train is received by one input neuron of the spiking neural network. With a predefined target pattern for each input pattern, the spiking neural network equipped with a supervised spike-timing based learning as described in Sect. 2.3 is trained to recognize the different spatiotemporal patterns

where $\tau = RC$ is the membrane time constant, $C = 1\,\mathrm{nF}$ is the membrane conductance, $R = 10\,\mathrm{M\Omega}$ is the membrane resistance, V is the membrane potential and $V_r = -60\,\mathrm{mV}$ is the rest potential, $I_0 = 0.1\,\mathrm{nA}$ is the constant inject current, I_{in} is the summation of presynaptic input currents, and I_n is a background noise modeled as a Gaussian process with zero mean and variance 1 nA. Once the membrane potential reaches the threshold $V_{thr} = -55\,\mathrm{mV}$, it will be reset to $V_{res} = -65\,\mathrm{mV}$ and held there for the refractory period.

The spike-based model presented here consists of two components: the latency-phase encoding and the supervised spike-timing based learning. Starting from environmental stimuli, we first encode images into spatiotemporal patterns and then transmit them to a spiking neural network for learning. The entire structure of the model is illustrated in Fig. 3.1.

3.2.2   Latency-Phase Encoding

With a combination of temporal encoding and phase encoding, a feature-dependent phase encoding algorithm has been proposed in [38]. Inspired by the information processing in the retina, the visual information is encoded into the responses of neurons using precisely timed action potentials. The intensity value of each pixel is converted to a precisely timed spike via a latency encoding scheme. Various experiments show that a strong stimulation leads to a short spike latency, and a weak stimulation results in a long reaction time [39–41]. Therefore, a monotone decreasing function could be used for the conversion from sensory stimuli to spatiotemporal patterns. Here, a logarithmic intensity transformation is adopted, which is similar to that used in [42].

$$t_i = f(s_i) = t_{max} - ln(\alpha \cdot s_i + 1) \tag{3.2}$$

where t_i is the firing time of neuron i, t_{max} is the maximum time of encoding window, α is a scaling factor, and s_i is the intensity of the analog stimulation. One advantage of the logarithmic function is that the time differences of spike latencies are invariant with different contrast level, e.g., it depends on the relative strength of the stimulation.

Ganglion cells have been observed to be firing in synchrony in several species [43–45], which illustrates the involvement of oscillations in the retina. We assume that the phases of oscillations are related to action potentials and contribute to the information compression from photoreceptor cells to ganglion cells. To take advantage of the phase information, spikes are assigned with phases related to their respective oscillations. Since each ganglion cell receives spikes from a group of photoreceptor cells, which is defined as the receptive field of this ganglion cell, we assign different initial phases to their subthreshold membrane oscillations. The periodic oscillation is described as cosine function for simplicity,

$$i_{osc} = A \cos(\omega t + \phi_i) \tag{3.3}$$

where A is the magnitude of the subthreshold membrane oscillations, ω is the phase angular velocity of the oscillation, and ϕ_i is the phase shift of the ith neuron in the receptive field.

In order to distinguish photoreceptor cells in the same receptive field, we set a constant phase gradient among photoreceptor neurons. The phase of subthreshold membrane oscillation for the ith photoreceptor neuron ϕ_i is defined as:

$$\phi_i = \phi_0 + (i - 1) \cdot \Delta\phi \tag{3.4}$$

where ϕ_0 is the reference initial phase, and $\Delta\phi$ is the constant phase difference between nearby photoreceptor cells ($\Delta\phi < \frac{2\pi}{N_{RF}}$, N_{RF} is the number of photoreceptor cells in each receptive field).

The spikes generated by the photoreceptor cells in each receptive field are compressed into one spike train by the ganglion cell. In order to utilize the phase information of spikes to reconstruct the original visual stimuli, the alignment operation is required to link each spike in the spike train with the corresponding photoreceptor cell in the receptive field. The alignment procedure is implemented by forcing photoreceptor cells to fire only when the subthreshold membrane potentials reach their nearest peaks as illustrated in Fig. 3.2b, c. After compression as shown in Fig. 3.2c, d, each spike in the compressed spike train is linked to one particular photoreceptor cell in the receptive field according to the phase of the subthreshold oscillations. Consequently, the phase information and the alignment together build an one-to-one relationship between the photoreceptor cells and spikes generated by the corresponding ganglion cell. With the latency-phase coding scheme, external stimulation is encoded into precisely timed spikes and then compressed into spike trains. The intensity information is encoded into firing times while the spatial information is reserved by the phases of spikes. When the spike trains are transmitted to coupled networks with respect to the encoding area, latency-phase encoded spikes generated by photoreceptor cells can be reconstructed from the compressed spike trains with a same phase reference as shown in Fig. 3.2d, e. The visual stimulus can then be reconstructed via a simple latency decoding process as shown in Fig. 3.2e, f. The complete latency-phase scheme is illustrated in Fig. 3.2.

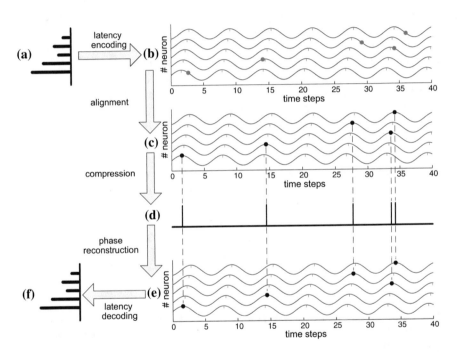

Fig. 3.2 Flowchart of the latency-phase encoding scheme. (**a**) Original stimuli. Stimulations with different intensities are the inputs to the photoreceptor cells. (**b**) The latency-encoded pattern. The visual information carried by the intensities is converted into the latencies of spikes. The spikes are assigned with phase information according to their corresponding oscillations. (**c**) Encoded spikes after latency encoding and alignment operation. The spikes are forced to be generated at peaks of the sub-threshold oscillations. (**d**) Compressed spike train. The spikes generated by the photoreceptor cells from the same receptive field are compressed into a spike train. (**e**) Reconstructed latency-encoded spikes. Spatial information within the receptive field could be retrieved from the compressed spike train via a phase reconstruction. (**f**) Decoded stimuli. By an inverse latency transformation, the original stimuli are reconstructed from the reconstructed spikes [38]

3.2.3 Supervised Spike-Timing Based Learning

It is known that learning from instructions is an important way for our brain to obtain new knowledge. As proposed in [36], the remote-supervised-method (ReSuMe) is compatible with temporal codes and is capable of performing spike-timing based learning precisely with millisecond timescale. The learning algorithm is based on a STDP-like process and synaptic modification during training depends on the pre- and postsynaptic firing times. After the training is successful, responses of output neurons will converge to the target patterns with a high time precision.

It is common that error signal between the target and the actual output is used in supervised learning. Similar to Widrow-Hoff rule applied in rate-based neuron models [46], the modification of synaptic efficacy in ReSuMe is triggered by either the target output ($S_d(t)$) or the actual output ($S_o(t)$). At the same time, the sign of error signal ($S_d(t) - S_o(t)$) decides the direction of the modification. To take the spike-timing into consideration, a STDP-like term is incorporated in the kernel a_{di}:

$$a_{di}(-s) = A \cdot exp(\frac{s}{\tau}), \quad \text{if } s < 0 \tag{3.5}$$

where A is the maximal magnitude of the STDP window, and s is the delay between the pre- and postsynaptic firing. Similar to the STDP process, if a presynaptic spike precedes a postsynaptic spike within a time interval, the synapse is strengthened. When the phase relation is reversed, the synapse is weaken. The magnitude of modification is determined by the lag s between pre- and postsynaptic spikes and is calculated by the convolution $a_{di}(t) * S_i(t)$. The complete learning rule is described as in Ponulak and Kasinski [36],

$$\frac{d}{dt}w_{oi}(t) = [S_d(t) - S_o(t)][a_d + \int_0^\infty a_{di}(s)S_i(t - s)ds] \tag{3.6}$$

where w_{oi} is the synaptic weight from the presynaptic neuron i to the postsynaptic neuron o. $S_d(t)$, $S_o(t)$ and $S_i(t)$ are the desired output, actual output and input spike train, respectively. a_d is a constant that helps speed up the learning process. From Eq. (3.6), we can see that the synaptic weights are updated when $S_d(t) \neq S_o(t)$, and the direction of modification is determined by the sign of the error signal $S_d(t) - S_o(t)$. No modification is induced when the actual output pattern is in agreement with the desired output pattern ($S_d(t) = S_o(t)$), which is used as the stopping criterion. The magnitude of modification is determined by the convolution term $a_{di}(t) * S_i(t)$. Thus, $S_i(t)$, $S_d(t)$ and $S_o(t)$ together are responsible for the synaptic modification. The learning rule is illustrated in Fig. 3.3.

The supervised signal is generated by the remote supervision scheme. Therefore, the target spike train is not directly delivered to the postsynaptic learning neuron and it determines the change of the synaptic efficacy from the presynaptic neuron

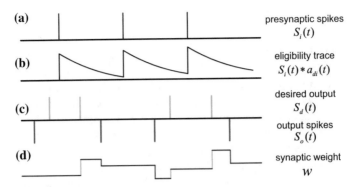

Fig. 3.3 Learning rule of ReSuMe. (**a**) The presynaptic input spikes, (**b**) The eligibility trace, (**c**) The desired output and actual output spikes, (**d**) The synaptic weight. The eligibility trace in (**b**) records the status of neuron according to the presynaptic spikes in (**a**). The desired output (positive direction) and the actual output (negative direction) in (**c**) together determine the sign of the supervisory signal. There is no other modification when the actual output spikes are generated at the desired times. The synaptic weight is updated when either a actual spike is generated or a desired spike should be induced. Meanwhile, the amount of synaptic weight change depends on the lag between pre- and postsynaptic spikes and the eligibility trace in (**b**) [36]

to postsynaptic neuron. It should be noted that both the excitatory synapses and inhibitory synapses exist in the model. During the learning, the synaptic weight is modified when either a target spike is needed or the postsynaptic learning neuron fires at the wrong time. When the modification occurs, the sign of error signal $(S_d(t) - S_o(t))$ decides the direction of change and the kernel $a_d + \int_0^\infty a_{di}(s)S_i(t-s)ds$ decides the amount of weight change. The synapses contributing to the firing of desired spikes are excitatory and adjusted to bring forward or hold off the firing times. On the other hand, the inhibitory synapses are used to suppress the firings at undesired times. The learning process stops as soon as the actual output patterns are identical to the target patterns.

3.3 Numerical Simulations

Real-world visual stimuli are often complex and contain a large amount of information. In this section, three 256×256 grayscale images are used to demonstrate the classification capability and the robustness of the integrated model. Images from the Urban and Natural Scene Categories of the LabelMe data set [47] are used here to explore the influence of parameter variations and the memory capacity of the system.

3.3.1 Network Architecture and Encoding of Grayscale Images

The receptive field (RF) of a sensory neuron is defined as a spatial region where the presence of stimulus affects the firing of that neuron. During the encoding phase, visual information from photoreceptor cells in the same RF is projected to retinal ganglion cells. Each ganglion cell then compresses the received spikes into a spike train. Therefore, the number of spikes in each spike train is determined by the number of pixels in each input image and the number of RFs.

$$N_{spike} = \frac{n}{N_{RF}} \tag{3.7}$$

where N_{spike} is the number of spikes in each spike train (number of pixels in each sub-field assigned with an RF), n is the number of photoreceptor cells (number of pixels of each image), and N_{RF} is the number of retinal ganglion cells (i.e., the number of RFs). Since each ganglion cell connects to one input neuron of the consecutive spiking neural network, the number of input neurons N is equal to N_{RF}. The number of output neurons depends on the size of data sets and the readout strategy. Intuitively, for large database with a large number of classes and complex target patterns with more spikes, more output neurons are required to perform the learning task. A two layer spiking neural network with 1024 input neurons and a single output neuron is used to illustrate the recognition capability of this model.

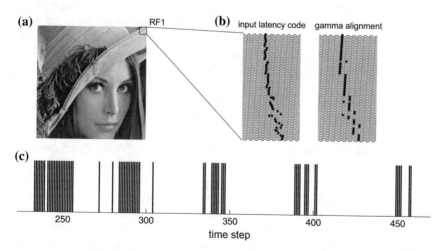

Fig. 3.4 The latency-phase encoding. The original image (256×256 pixels) in (**a**) is partitioned into 1024 RFs with the size of 8×8. The left pattern in (**b**) is the spike pattern of RF1 after latency encoding and the right one is the pattern further processed by the alignment operation (spikes are denoted by the dot markers). The compressed spike train of RF1 is given in (**c**). For better visualization, only part of the encoded spatiotemporal pattern is illustrated

Here, grayscale images with the size of 256×256 pixels are used as the external stimulation. Each pixel value is regarded as the intensity of the visual stimulation received by the photoreceptor cell in the retina. Thus there are 1024 RFs with the size of 8×8 pixels as shown in Fig. 3.4a. After the alignment as shown in Fig. 3.4b, each ganglion cell receives 64 spikes from 64 photoreceptor cells in its receptive field and compresses them into one spike train as shown in Fig. 3.4c. Therefore, information of the 256×256 pixel image is encoded into 1024 spike trains and each spike train contains 64 spikes. As the encoding method converts the intensity values into firing times of spikes, the visual information is preserved by the temporal configuration of the spike trains.

3.3.2   Learning Performance

To recognize images, we predefine different target spike patterns for input patterns. For simplicity, each target pattern is defined as a sequence of three spikes (each target pattern is denoted by a different marker type, as shown in Fig. 3.5a). After sensory encoding, three spatiotemporal patterns of length 640 ms are repetitively presented to the network in a random sequence. The number of epoch is increased when one pattern has been presented to the network, while the number of iteration is increased when all patterns have been presented to the network once. The responses of the output neuron for different input patterns are shown in Fig. 3.5a. To quantitively evaluate the learning performance, a correlation-based measure of spike timing [48] is adopted to measure the distance between the output pattern and the target pattern. The correlation C is close to unity when the output pattern matches the target pattern and equals to zero when the two patterns are unrelated. The spike trains (S_o and S_d) are convolved with a low pass Gaussian filter of a given width $\sigma = 2$ ms. If the filtered spike trains are $\vec{s_1}$ and $\vec{s_2}$, the correlation measure is

$$C = \frac{\vec{s_1} \cdot \vec{s_2}}{|\vec{s_1}||\vec{s_2}|} \tag{3.8}$$

The typical results of the training are shown in Fig. 3.5. Within 20 presentations of each input pattern, the output neuron is able to reproduce the target pattern as shown in Fig. 3.5. At first, the output neuron fires at random times. After several iterations, extra spikes firing at undesired times disappear, and the actual output patterns approach to the corresponding target patterns. When successful learning is achieved, the output neuron is able to reproduce different target patterns when different input patterns are given. We repeated the training for dozens of times and observed that the spiking neuron is able to learn the training pairs successfully.

3.3.3 Generalization Capability

The integrated model recognizes each image as a certain spatiotemporal pattern, in which the intensities of individual pixels are encoded into precisely timed spikes. Therefore, the generalization of the system is expected to be related to the pixel-level features of the input images. To study the generalization capability of the model, we add different levels of Gaussian, speckle and salt-and-pepper noise to the input images during the testing phase. The Gaussian noise is specified by its mean m and variance v, the speckle noise is specified by its variance v, and the salt-and-pepper noise is specified by the noise density d. For each kind of the noise with different intensities, we test the trained network with one hundred noisy images. The test

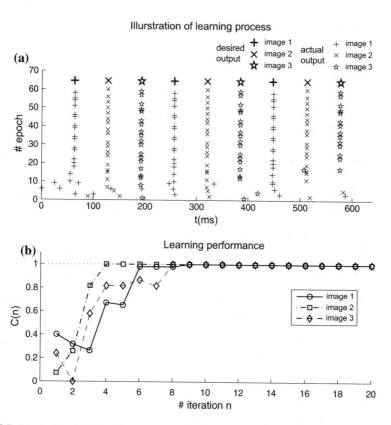

Fig. 3.5 Illustration of the learning process and performance. (**a**) Raster plot of the output spikes. When presented with different input patterns, the output patterns converge to the corresponding target patterns. Given different input patterns, spikes generated by the output neuron are denoted by different marker types. (**b**) The correlations C between output spike trains and the target spike trains against learning iterations. At first, the output neuron fires at random times. After several iterations, the output patterns begin to approach to the target patterns and the learning is converged within twenty iterations

results are shown in Fig. 3.6b. By analyzing the learning process, we can see that the pixel-feature dependent generalization is related to temporally local learning algorithm. During the learning process, only the synaptic weights associated with input spikes evoking the postsynaptic spikes within the learning window are updated. The decaying learning window makes the optimization process to be focused on a limited number of synapses, which affects the firing time of the nearest postsynaptic neuron. At the same time, noise added to input images shifts part of the firing times of the encoded spatiotemporal pattern. Therefore, the spiking neuron should be able to reproduce target spikes with a small temporal error in response to the input images with pixel noise, but fail to recognize images in the presence of other type of noises. As expected, the test results in Fig. 3.6b show that the system is more resistant to salt-and-pepper noise than speckle noise or Gaussian noise.

We also add the different type of noises to the input images during the training phase. For each type of noise, 100×3 noisy images are used as the training set. After

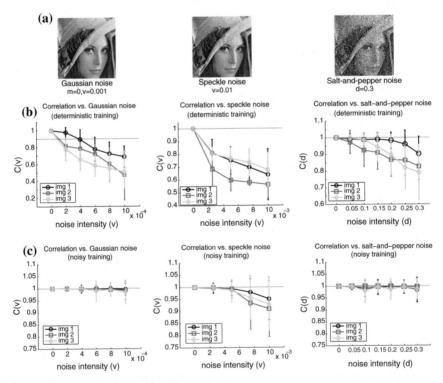

Fig. 3.6 The test results with different type of noises added to the input images. (**a**) Examples of images with different type of noises, such as Gaussian, speckle and salt-and-pepper noise. The correlation C between the output spike pattern and the target pattern is used to evaluate the precision of the neural responses. (**b**) Reliable responses can be reproduced by the spiking neural networks for noisy images (e.g., deterministic training). (**c**) The robustness to noise is improved when the noise information is included during the training phase (e.g., noisy training)

training, another 100×3 images with noise of the same type and intensity level are used to examine the reliability of the neural responses after noisy training. As shown in Fig. 3.6c, when the noise information is learned by the classifier during training phase, the robustness of the system due to the effect of noise has been improved. It can also be observed that the maximum level of salt-and-pepper noise that the system can tolerate is much higher than that of the other two type of noises, which is consistent with our analysis.

3.3.4 Parameters Evaluation

To examine the influence of parameter variations in the encoded patterns, 100 images (256×256 pixel, 8-bit grayscale) from the Urban and Natural Scene Categories of the LabelMe database are encoded with various parameter configurations. The images from LabelMe data set are used here to study the properties of the integrated model due to their distributed intensity values and their closeness to real-world stimulation. A few sample images from the data set are given in Fig. 3.7.

The size of receptive field, encoding cycles and phase shift constant are important parameters for the encoding method. Since photoreceptor cells of the same RF convey visual information to the corresponding retinal ganglion cell, the number of photoreceptor cells in each RF affects the number of spikes in the compressed spike train. If the length of encoding window is fixed, increasing the RF size would result in a higher average firing rate of the compressed spike trains.

Fig. 3.7 Sample images of "buildings inside city" category from the LabelMe database. The original 256×256 color images are converted into 8-bit grayscale images

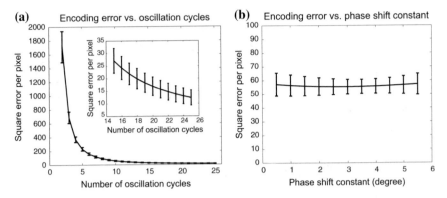

Fig. 3.8 The encoding error with different encoding cycles and phase shift constants on natural images from the LabelMe database. The average square error per pixel (vertical axis) is employed to estimate the encoding accuracy of the test images. (**a**) The encoding error drops when the number of oscillation cycles increases. With more subthreshold membrane oscillation cycles, more oscillation peaks provide more sampling points to encode input intensities (the tail of the curve is enlarged in the inset). (**b**) The phase shift constant $\Delta\phi$ slightly affects the encoding accuracy

Considering the accuracy of encoding process, no error is introduced by the latency encoding scheme. The distortion of information is resulted from the alignment operation. As the alignment operation moves spikes to the peaks of the subthreshold oscillations, the encoding accuracy is affected by the number of oscillation cycles within the encoding period as shown in Fig. 3.8a. To estimate the accuracy of encoding, we compare the reconstructed images with the original images using the average square of error per pixel,

$$e = \frac{\sum_{i=1}^{n} (s_i - s_i')^2}{n} \tag{3.9}$$

where s_i and s_i' are the intensities of the ith pixel in the original image and the reconstructed image, respectively.

Since the intensity information is carried by the temporal spikes, the distribution of the original images as well as the encoding parameters such as phase shift resolution $\Delta\phi$ may affect the temporal distribution of the encoded spatiotemporal patterns. The experiment results illustrate that the phase shift constant hardly affects the encoding accuracy as shown in Fig. 3.8b. However, it will determine the spike distribution of the compressed spike train as shown in Fig. 3.9. The encoded spikes concentrate in the time domain with a small shift constant as shown in Fig. 3.9a and spread out with a large shift constant as shown in Fig. 3.9b.

Therefore, the choice of encoding cycles depends on the precision requirement for a specific application. Since the phase shift resolution $\Delta\phi$ affects the distribution of encoded spatiotemporal patterns, it should be tuned according to the learning algorithm adopted in the posterior neural network.

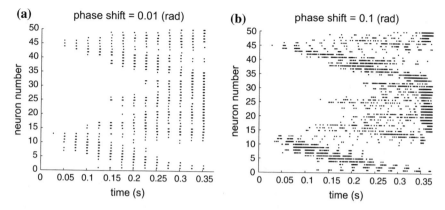

Fig. 3.9 The encoded patterns with a different phase shift constant. The phase shift constant is the phase difference between nearby photoreceptor cells in the same receptive field and affects the firing times within each receptive field. With a small phase shift constant, neurons within the same receptive field tend to fire simultaneously as shown in (**a**). With a large phase shift constant, the temporal distribution of spikes is scattered as shown in (**b**)

Since the postsynaptic depolarization is determined by the integration of presynaptic input spikes, temporal distribution of input spatiotemporal patterns and the complexity of target patterns will affect the learning performance. On one hand, because a target spike requires one or more preceding input spikes to excite the output neuron to fire at the desired time, enough presynaptic input spikes are needed for the generation of spikes. On the other hand, increasing the number of target spikes will result in competition for limited available synapses between the target spikes firing at different times and impose restriction on the behavior of the output neuron. We tested the system on 100 images (128 × 128 8-bit grayscale images from Urban and Natural Scene Categories of LabelMe database) to examine the influence of target patterns on the learning performance. For each number of target spikes, the network was trained with one randomly generated target pattern. It is observed that the spiking neuron needs more iterations to achieve a successful learning for a more complex target patterns as discussed in our analysis.

3.3.5 Capacity of the Integrated System

The spiking neural network with the same settings in previous experiments is used to explore the memory capacity of the integrated system. From a computational point of view, precisely timed spikes have a remarkable encoding capacity, i.e., the memory capacity of the system is often limited by the learning scheme employed. Since most of the information is reserved by the temporal code, the design of target patterns plays a pivot role in exploiting the information carried by the encoded spatiotemporal

Fig. 3.10 Memory (or recognition) capacity of the integrated model. The average percentage of successful recall of patterns is plotted as a function of training pairs. The successful recall percentage drops dramatically after the number of training pairs is larger than 11

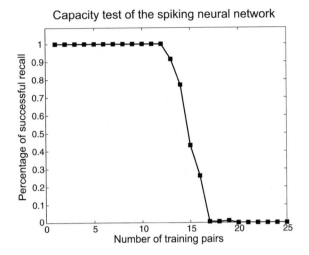

patterns. We train the network with different number of input patterns and define the percentage of successful recall of trained pairs as an evaluation of the memory capacity. A successful recall of one trained pattern is achieved when the distance between the output pattern of the trained network and the target pattern is close enough, i.e., $C > 0.95$ as the threshold. To simplify the problem for a classification task, we randomly generated one target spike train containing ten spikes for all input images every time and repeat the experiment for 20 times.

As shown in Fig. 3.10, for the 1024-1 spiking neural network with ten spikes in the target patterns and the selected parameter settings, around 11 training pairs can be successfully stored and recalled with a slight time shift. The percentage of successful recall decreases quickly when the number of training pairs is increased. Apparently, it can be inferred that decreasing the number of target spikes (complexity) or increasing the free tunable parameters will lead to a larger amount of information capacity. However, this would also allow less information of the spatiotemporal patterns to be learned. Although it is not mathematically analyzed, the presented simulation results for the specific case provide some insight into the information capacity of the system.

To summarize it from a system level, temporally distributed input spatiotemporal patterns and simple target patterns are preferred for better generalization capabilities and memory capacity of the integrated model. The scattered distribution of input patterns enables the output neuron to generate spikes at arbitrary times. Although the network can learn more about the original images with more complex target patterns, the computational efforts will also be increased and the information capacity will be limited. Therefore, the tradeoff between the learning level of input patterns and the computational efforts as well as memory capacity should be considered for any specific applications.

3.4 Related Works

Spiking neural networks have been applied to solve different classification tasks [31, 49–52]. Hopfield and Brody [30] proposed a computational model for pattern recognition, in which analog signal is employed as neural representation of sensory stimuli. The transient synchronization of decaying delay activity of a specific subset of input neurons are used for recognition. Although it has been successfully applied to speech recognition [31] and olfactory recognition [49], the unknown mechanism of encoding input stimulation into decay firing activities makes the model questionable. Bohte et al. [50] proposed a temporal version of error-backpropagation, SpikeProp. The SpikeProp was demonstrated to be able to classify images with a three-layer spiking neural network. However, the adaptive learning can only be applied to analytically tractable neuron models, and the weights with mixed signs are suspected to cause failures of training [53]. Gütig and Sompolinsky [51] proposed a supervised learning algorithm, temptron, to classify spatiotemporal patterns by generating at least one spike or staying quiescent.

Brader, Senn and Fusi [52] proposed an alternative approach, in which a spike-driven model is able to perform binary image classification with spiking neurons using rate codes. In this approach, grayscale value of each pixel of input images is normalized to a binary value such that the largest element is unity. Then each element was encoded by Poisson spike trains at different frequencies. After learning, images from different classes can be distinguished by the firing rates of output neurons. However, the spike-driven model only focuses on the learning part and pay little attention to the sensory encoding. By transforming 8-bit grayscale images into binary images, a large amount of the images have been discarded. Therefore, the actual information carried by the input patterns are far less than that of the original images. Moreover, the spike-driven learning relies on a stochastic process, which makes the learning algorithm less efficient and computational demanding.

Due to the use of different encoding scheme and learning strategy, the proposed integrated model has several advantages over existing approaches. First, we look at the pattern recognition process at a system level. Rather than considering sensory encoding and learning as isolated processes, we integrated biological plausible encoding and learning processes using consistent neural codes. The latency-phase encoding scheme retains almost all information of the input images with high precision and links up the sensory encoding with learning process. Second, in the integrated spike-based model, we demonstrated that input patterns can be classified by precisely timed spike trains rather than the mean firing rates or single spike code. With the rich capacity of temporal codes, detailed information of the inputs can be exploited by designing the target pattern and precisely timed spikes can be generated. Furthermore, the supervised spike-timing based learning allows an efficient computation and fast convergence, such that the system can be applied to real-life tasks, such as movement control [54] and neuroprostheses control [55].

The input neurons are supposed to fire more than once in our model, which makes better use of the synaptic weights and generalization performance. Although the temporal codes provide a large amount of information, multi-spike signal results in the competition among target spikes firing at different times for the available resources. This leads to limited memory capacity and slow convergence as shown in the simulation results. Therefore the removal of the conflicts among the target spikes remains a challenging but interesting issue for the spike-timing based learning algorithm. One approach is to employ multiple layer and recurrent neural structures, such as liquid state machine [56], so as to increase the computational capability of the system and to absorb the influence of multiple spikes.

There are a few limitations in our current model. The encoding scheme in the model does not incorporate any information extraction to preprocess the input patterns, which is viewed as a necessary procedure in traditional pattern recognition models. By using filtering techniques as proposed in HMAX model [57] or local edge detectors [58], it is believed that the performance and memory capacity in the proposed model will be improved with an efficient neural code in a more concise and abstract manner.

3.5   Conclusions

In this chapter, an integrated computational model with latency-phase encoding method and supervised spiking-timing based learning algorithm has been proposed. Stimuli were first encoded into spatiotemporal patterns with latency-phase scheme, which builds up a bridge between real-world stimuli to neural signals in a biological plausible way. Then the patterns were learned by spiking neurons using a spike-timing based supervised method with millisecond time precision. As shown in the simulation results, the spike-timing based neural networks with temporal codes are capable of solving pattern recognition task by computing with action potentials.

Although the current model has limitations in the recognition capacity, our study exploits the computational mechanisms employed by neural systems in two respects: First, our model was built at a system level emphasizing both the sensory encoding and learning process. It is an integrated system based on a unified temporal coding scheme and consistent with the known neurobiological mechanisms. Second, we have demonstrated the classification capability of the system that computes precisely timed spikes and realistic stimuli, analogously to cognitive computation in human brain. The approaches based on cognitive computation will play a leading role in many applications spanning across signal processing, autonomous systems and robotics [59–61].

References

1. Du Bois-Reymond, E.: Untersuchungen uer thierische elektricita. G. Reimer (1848)
2. Panzeri, S., Brunel, N., Logothetis, N.K., Kayser, C.: Sensory neural codes using multiplexed temporal scales. Trends Neurosci. **33**(3), 111–120 (2010)
3. Softky, W.R.: Simple codes versus efficient codes. Curr. Opin. Neurobiol. **5**(2), 239–247 (1995)
4. Rullen, R.V., Thorpe, S.J.: Rate coding versus temporal order coding: what the retinal ganglion cells tell the visual cortex. Neural Comput. **13**(6), 1255–1283 (2001)
5. Adrian, E.: The Basis of Sensation: The Action of the Sense Organs. W. W. Norton, New York (1928)
6. Shadlen, M.N., Newsome, W.T.: Noise, neural codes and cortical organization. Curr. Opin. Neurobiol. **4**(4), 569–579 (1994)
7. Litvak, V., Sompolinsky, H., Segev, I., Abeles, M.: On the transmission of rate code in long feedforward networks with excitatory-inhibitory balance. J. Neurosci. **23**(7), 3006–3015 (2003)
8. Seung, H.S.: Learning in spiking neural networks by reinforcement of stochastic synaptic transmission. Neuron **40**(6), 1063–1073 (2003)
9. Barak, O., Tsodyks, M.: Recognition by variance: learning rules for spatiotemporal patterns. Neural Comput. **18**, 2343–2358 (2006)
10. Bialek, W., Rieke, F., de Ruyter van Steveninck, R., Warland, D.: Reading a neural code. Science **252**(5014), 1854–1857 (1991)
11. Victor, J.D.: How the brain uses time to represent and process visual information. Brain Res. **886**(1–2), 33–46 (2000)
12. Carr, C.E.: Processing of temporal information in the brain. Annu. Rev. Neurosci. **16**(1), 223–243 (1993)
13. Singer, W., Gray, C.M.: Visual feature integration and the temporal correlation hypothesis. Annu. Rev. Neurosci. **18**(1), 555–586 (1995)
14. Kayser, C., Montemurro, M.A., Logothetis, N.K., Panzeri, S.: Spike-phase coding boosts and stabilizes information carried by spatial and temporal spike patterns. Neuron **61**(4), 597–608 (2009)
15. Meister, M., II, M.J.B.: The neural code of the retina. Neuron **22**(3), 435–450 (1999)
16. Gollisch, T., Meister, M.: Rapid neural coding in the retina with relative spike latencies. Science **319**(5866), 1108–1111 (2008)
17. Keat, J., Reinagel, P., Reid, R., Meister, M.: Predicting every spike: A model for the responses of visual neurons. Neuron **30**(3), 803–817 (2001)
18. Llinas, R.R., Grace, A.A., Yarom, Y.: In vitro neurons in mammalian cortical layer 4 exhibit intrinsic oscillatory activity in the 10-to 50-Hz frequency range. Proc. Natl. Acad. Sci. **88**(3), 897–901 (1991)
19. Koepsell, K., Wang, X., Vaingankar, V., Wei, Y., Wang, Q., Rathbun, D.L., Usrey, W.M., Hirsch, J.A., Sommer, F.T.: Retinal oscillations carry visual information to cortex. Front. Syst. Neurosci. **3**, 4 (2009)
20. Heiligenberg, W.: Neural Nets in Electric Fish. MIT Press, Cambridge (1991)
21. Chrobak, J.J., Buzsáki, G.: Gamma oscillations in the entorhinal cortex of the freely behaving rat. J. Neurosci. **18**(1), 388–398 (1998)
22. O'Keefe, J., Burgess, N.: Dual phase and rate coding in hippocampal place cells: theoretical significance and relationship to entorhinal grid cells. Hippocampus **15**(7), 853–866 (2005)
23. Tsodyks, M.V., Skaggs, W.E., Sejnowski, T.J., McNaughton, B.L.: Population dynamics and theta rhythm phase precession of hippocampal place cell firing: a spiking neuron model. Hippocampus **6**(3), 271–280 (1996)
24. Jensen, O.: Information transfer between rhythmically coupled networks: reading the hippocampal phase code. Neural Comput. **13**(12), 2743–2761 (2001)
25. Blumenfeld, B., Preminger, S., Sagi, D., Tsodyks, M.: Dynamics of memory representations in networks with novelty-facilitated synaptic plasticity. Neuron **52**(2), 383–394 (2006)
26. Tang, H., Li, H., Yan, R.: Memory dynamics in attractor networks with saliency weights. Neural Comput. **22**(7), 1899–1926 (2010)

27. Mainen, Z., Sejnowski, T.: Reliability of spike timing in neocortical neurons. Science **268**(5216), 1503–1506 (1995)
28. Bi, G.Q., Poo, M.M.: Synaptic modifications in cultured hippocampal neurons: dependence on spike timing, synaptic strength, and postsynaptic cell type. J. Neurosci. **18**(24), 10464–10472 (1998)
29. Bi, G.Q., Poo, M.M.: Synaptic modification by correlated activity: Hebb's postulate revisited. Annu. Rev. Neurosci. **24**, 139–166 (2001)
30. Hopfield, J.J., Brody, C.D.: What is a moment? "cortical" sensory integration over a brief interval. Proc. Natl. Acad. Sci. **97**(25), 13919–13924 (2000)
31. Hopfield, J.J., Brody, C.D.: What is a moment? transient synchrony as a collective mechanism for spatiotemporal integration. Proc. Natl. Acad. Sci. **98**(3), 1282–1287 (2001)
32. Ito, M.: Mechanisms of motor learning in the cerebellum. Brain Res. **886**(1–2), 237–245 (2000)
33. Montgomery, J., Carton, G., Bodznick, D.: Error-driven motor learning in fish. Biol. Bull. **203**(2), 238–239 (2002)
34. Knudsen, E.I.: Supervised learning in the brain. J. Neurosci. **14**(7), 3985–3997 (1994)
35. Ito, M.: Control of mental activities by internal models in the cerebellum. Nat. Rev. Neurosci. **9**(4), 304–313 (2008)
36. Ponulak, F., Kasinski, A.: Supervised learning in spiking neural networks with resume: sequence learning, classification, and spike shifting. Neural Comput. **22**(2), 467–510 (2010)
37. Gerstner, W., Kistler, W.M.: Spiking Neuron Models: Single Neurons, Populations, Plasticity. Cambridge University Press, Cambridge (2002)
38. Nadasdy, Z.: Information encoding and reconstruction from the phase of action potentials. Front. Syst. Neurosci. **3**, 6 (2009)
39. Gawne, T.J., Kjaer, T.W., Richmond, B.J.: Latency: another potential code for feature binding in striate cortex. J. Neurophysiol. **76**(2), 1356–1360 (1996)
40. Reich, D.S., Mechler, F., Victor, J.D.: Temporal coding of contrast in primary visual cortex: when, what, and why. J. Neurophysiol. **85**(3), 1039–1050 (2001)
41. Greschner, M., Thiel, A., Kretzberg, J., Ammermüller, J.: Complex spike-event pattern of transient on-off retinal ganglion cells. J. Neurophysiol. **96**(6), 2845–2856 (2006)
42. Hopfield, J.J.: Pattern recognition computation using action potential timing for stimulus representation. Nature **376**(6535), 33–36 (1995)
43. Arnett, D.: Statistical dependence between neighboring retinal ganglion cells in goldfish. Exp. Brain. Res. **32**(1) (1978)
44. DeVries, S.H.: Correlated firing in rabbit retinal ganglion cells. J. Neurophysiol. **81**(2), 908–920 (1999)
45. Meister, M., Lagnado, L., Baylor, D.A.: Concerted signaling by retinal ganglion cells. Science **270**(5239), 1207–1210 (1995)
46. Widrow, B., Hoff, M.E., et al.: Adaptive switching circuits (1960)
47. Russell, B.C., Torralba, A., Murphy, K.P., Freeman, W.T.: Labelme: a database and web-based tool for image annotation. Int. J. Comput. Vis. **77**(1–3), 157–173 (2008)
48. Schreiber, S., Fellous, J., Whitmer, D., Tiesinga, P., Sejnowski, T.: A new correlation-based measure of spike timing reliability. Neurocomputing **52–54**, 925–931 (2003)
49. Brody, C.D., Hopfield, J.: Simple networks for spike-timing-based computation, with application to olfactory processing. Neuron **37**(5), 843–852 (2003)
50. Bohte, S.M., Bohte, E.M., Poutr, H.L., Kok, J.N.: Unsupervised clustering with spiking neurons by sparse temporal coding and multi-layer RBF networks. IEEE Trans. Neural Netw. **13**, 426–435 (2002)
51. Gütig, R., Sompolinsky, H.: The tempotron: a neuron that learns spike timing-based decisions. Nat. Neurosci. **9**(3), 420–428 (2006)
52. Brader, J.M., Senn, W., Fusi, S.: Learning real-world stimuli in a neural network with spike-driven synaptic dynamics. Neural Comput. **19**(11), 2881–2912 (2007)
53. Haruhiko, T., Masaru, F., Hiroharu, K., Shinji, T., Hidehiko, K., Terumine, H.: Obstacle to training spikeprop networks: cause of surges in training process. In: Proceedings of the 2009 International Joint Conference on Neural Networks, pp. 1225–1229. IEEE Press, Piscataway (2009)

54. Manette, O., Maier, M.: Temporal processing in primate motor control: relation between cortical and EMG activity. IEEE Trans. Neural Netw. **15**(5), 1260–1267 (2004)
55. Müller-Putz, G.R., Scherer, R., Pfurtscheller, G., Neuper, C.: Temporal coding of brain patterns for direct limb control in humans. Front. Neurosci. **4** (2010)
56. Maass, W., Natschläger, T., Markram, H.: Real-time computing without stable states: a new framework for neural computation based on perturbations. Neural Comput. **14**(11), 2531–2560 (2002)
57. Riesenhuber, M., Poggio, T.: Hierarchical models of object recognition in cortex. Nature Nurosci. **2**(11), 1019–1025 (1999)
58. van Wyk, M., Taylor, W.R., Vaney, D.I.: Local edge detectors: a substrate for fine spatial vision at low temporal frequencies in rabbit retina. J. Neurosci. **26**(51), 13250–13263 (2006)
59. Perlovsky, L.: Computational intelligence applications for defense [research frontier]. Comput. Intell. Mag. IEEE **6**(1), 20–29 (2011)
60. Meng, Y., Zhang, Y., Jin, Y.: Autonomous self-reconfiguration of modular robots by evolving a hierarchical mechanochemical model. Comput. Intell. Mag. IEEE **6**(1), 43–54 (2011)
61. Yan, R., Tee, K.P., Chua, Y., Li, H., Tang, H.: Gesture recognition based on localist attractor networks with application to robot control [application notes]. Comput. Intell. Mag. IEEE **7**(1), 64–74 (2012)

Chapter 4
Precise-Spike-Driven Synaptic Plasticity for Hetero Association of Spatiotemporal Spike Patterns

Abstract This chapter introduces a new temporal learning rule, namely the Precise-Spike-Driven (PSD) Synaptic Plasticity, for processing and memorizing spatiotemporal patterns. PSD is a supervised learning rule that is analytically derived from the traditional Widrow-Hoff (WH) rule and can be used to train neurons to associate an input spatiotemporal spike pattern with a desired spike train. Synaptic adaptation is driven by the error between the desired and the actual output spikes, with positive errors causing long-term potentiation and negative errors causing long-term depression. The amount of modification is proportional to an eligibility trace that is triggered by afferent spikes. The PSD rule is both computationally efficient and biologically plausible. The properties of this learning rule are investigated extensively through experimental simulations, including its learning performance, its generality to different neuron models, its robustness against noisy conditions, its memory capacity, and the effects of its learning parameters.

4.1 Introduction

With the same capability of processing spikes as biological neural systems, SNNs [1–3] are more biologically realistic and computationally powerful than the traditional artificial neural networks. Spikes are believed to be the principal feature in the information processing of neural systems, though the neural coding mechanism, i.e., how information is encoded in spikes still remains unclear. The temporal codes describe one possibility, where information is conveyed through precise times of spikes. However, the complexity of processing temporal codes [4, 5] might limit their usage in SNNs, which demands the development of efficient learning algorithms.

Supervised learning was proposed as a successful concept of information processing [6]. Neurons are driven to respond at desired states under a supervisory signal, and an increasing body of evidence shows that this kind of learning is exploited by the brain [7–10]. Supervised mechanism has been widely used to develop various learning algorithms for processing spatiotemporal spike patterns in SNNs [5, 11–16].

© Springer International Publishing AG 2017
Q. Yu et al., *Neuromorphic Cognitive Systems*, Intelligent Systems Reference Library 126, DOI 10.1007/978-3-319-55310-8_4

SpikeProb [12] is one of the first supervised learning algorithms for processing precise spatiotemporal patterns in SNNs. However, in its original form, SpikeProb cannot learn to reproduce a multi-spike train. The tempotron rule [5], another gradient descent approach that is evaluated to be efficient for binary temporal classification tasks, cannot output multiple spikes either. As the tempotron is designed mainly for pattern recognition, it is unable to produce precise spikes. To produce a desired spike train, several learning algorithms have been proposed such as ReSuMe [13, 17], Chronotron [14] and SPAN [15]. These three learning rules are all capable of training a neuron to generate a desired spike train in response to an input stimulus. In the Chronotron, two learning rules are introduced. One is analytically-derived (E-learning) and another one is heuristically-defined (I-learning). The I-learning rule is more biologically plausible but comes with less memory capacity than the E-learning rule. The performance of the I-learning rule depends on the weight initialization, where initial zero values can cause information loss from the corresponding afferent neurons. The E-learning rule and the SPAN rule are both based on an error function of the difference between the actual output spike train and the desired spike train. Their applicability is therefore limited to the tractable error evaluation, which might be unavailable in actual biological networks and inefficient from a computational point of view. These arithmetic-based rules can reveal explicitly how SNNs can be trained but the biological plausibility of the error calculation is somewhat questionable.

In this chapter, we propose an alternative learning mechanism called Precise-Spike-Driven (PSD) synaptic plasticity, that is able to learn the association between precise spike patterns. Similar to ReSuMe [13] and SPAN [15], the PSD rule is derived from the Widrow-Hoff (WH) rule but based on a different interpretation. The PSD rule is derived analytically based on converting the spike trains into analog signals by applying the spike convolution method. Such an approach is rarely reported in the existing learning rule studies [15]. Synaptic adaptation in the PSD is driven by the error between the desired and the actual output spikes, with positive errors causing long-term potentiation (LTP) and negative errors causing long-term depression (LTD). The amount of adaptation depends on an eligibility trace determined by the afferent spikes. Without complex error calculation, the PSD rule provides an efficient way for processing spatiotemporal patterns. We show that the PSD rule inherits the advantageous properties of both arithmetic-based and biologically realistic rules, being simple and efficient for computation, and yet biologically plausible. Furthermore, the PSD is an independent plasticity rule that can be applied to different neuron models. This straightforward interpretation of the WH rule also provides a possible direction for further exploitation of the rich theory of ANNs, and minimizes the gap between the learning algorithms of SNNs and the traditional ANNs.

Various properties of the PSD rule are investigated through an extensive experimental analysis. In the first experiment, the basic concepts of the PSD rule are demonstrated, and its learning ability on hetero-association of spatiotemporal spike pattern is investigated. In the second experiment, the PSD rule is shown to be applicable to different neuron models. Thereafter, experiments are conducted to analyze the learning rule regarding its robustness against noisy conditions, its memory capacity, effects of the learning parameters and its classification performance. Finally, a detailed discussion about the PSD rule and several related algorithms is presented.

4.2 Methods

In this section, we begin by presenting the spiking neuron models. We then describe the PSD rule for learning hetero-association between the input spatiotemporal spike patterns and the desired spike trains.

4.2.1 Spiking Neuron Model

The LIF model is firstly considered. The dynamics of each neuron evolves according to the following equation:

$$\tau_m \frac{dV_m}{dt} = -(V_m - E) + (I_{ns} + I_{syn}) \cdot R_m \tag{4.1}$$

where V_m is the membrane potential, $\tau_m = R_m C_m$ is the membrane time constant, $R_m = 1\,M\Omega$ and $C_m = 10\,nF$ are the membrane resistance and capacitance, respectively, E is the resting potential, I_{ns} and I_{syn} are the background current noise and synaptic current, respectively. When V_m exceeds a constant threshold V_{thr}, the neuron is said to fire, and V_m is reset to V_{reset} for a refractory period t_{ref}. We set $E = V_{reset} = 0\,mV$ and $V_{thr} = E + 18\,mV$ for clarity, but any other values as $E = -60\,mV$ will result in equivalent dynamics as long as the relationships among E, V_{reset} and V_{thr} are kept.

For the post-synaptic neuron, we model the input synaptic current as:

$$I_{syn}(t) = \sum_i w_i I_{PSC}^i(t) \tag{4.2}$$

where w_i is the synaptic efficacy of the i-th afferent neuron, and I_{PSC}^i is the unweighted postsynaptic current from the corresponding afferent.

$$I_{PSC}^i(t) = \sum_{t^j} K(t - t^j) H(t - t^j) \tag{4.3}$$

where t^j is the time of the j-th spike emitted from the i-th afferent neuron, $H(t)$ refers to the Heaviside function, K denotes a normalized kernel and we choose it as:

$$K(t - t^j) = V_0 \cdot \left(\exp(\frac{-(t - t^j)}{\tau_s}) - \exp(\frac{-(t - t^j)}{\tau_f}) \right) \tag{4.4}$$

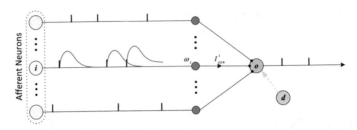

Fig. 4.1 Illustration of the neuron structure. The afferent neurons are connected to the post-synaptic neuron through synapses. Each emitted spike from afferent neurons will trigger a post-synaptic current (PSC). The membrane potential of the post-synaptic neuron is a weighted sum of all incoming PSCs from all afferent neurons. The *yellow* neuron denotes the instructor which is used for learning

where V_0 is a normalization factor such that the maximum value of the kernel is 1, τ_s and τ_f are the slow and fast decay constants respectively, and their ratio is fixed at $\tau_s/\tau_f = 4$.

Figure 4.1 illustrates the neuron structure. Each spike from the afferent neuron will result in a post-synaptic current (PSC). The membrane potential of the post-synaptic neuron is a weighted sum of all incoming PSCs over all afferent neurons.

In addition to the LIF model, we also investigate the flexibility of the PSD rule to different neuron models. For this, we use the IM model [18], where the dynamics of the IM model is described as:

$$
\begin{cases}
dV_m/dt = 0.04V_m^2 + 5V_m + 140 - U + I_{syn} + I_{ns} \\
dU/dt = a(bV_m - U) \\
\text{if } V_m \geq 30\,mV, \\
\text{then } V_m \leftarrow c,\, U \leftarrow U + d
\end{cases}
\tag{4.5}
$$

where V_m again represents the membrane potential. U is the membrane recovery variable. The synaptic current (I_{syn}) is in the same form as described before, and I_{ns} again represents the background noise. The parameters $a = 0.02, b = 0.2, c = -65$ and $d = 8$ are chosen such that the neuron exhibits a regular spiking behavior which is the most typical behavior observed in cortex [18].

For computational efficiency, the LIF model is used in the following studies of this chapter, unless otherwise stated.

4.2.2 PSD Learning Rule

In this section we describe in detail the PSD learning rule. Note that the spiking neuron models were developed from the traditional neuron models. In a similar way, we develop the learning rule for spiking neurons from traditional algorithms. Inspired by [15], we derive the proposed rule from the common WH rule. The WH rule is described as:

$$\Delta w_i = \eta x_i (y_d - y_o) \tag{4.6}$$

where η is a positive constant referring to the learning rate, x_i, y_d and y_o refer to the input, the desired output and the actual output, respectively.

Note that because the WH rule was introduced for the traditional neuron models such as perceptron, the variables in the WH rule are regarded as real-valued vectors. In the case of spiking neurons, the input and output signals are described by the timing of spikes. Therefore, a direct implementation of the WH rule does not work for spiking neurons. This motivates the development of the PSD rule.

A spike train is defined as a sequence of impulses triggered by a particular neuron at its firing time. A spike train is expressed in the form of:

$$s(t) = \Sigma_f \delta(t - t^f) \tag{4.7}$$

where t^f is the f-th firing time, and $\delta(x)$ is the Dirac function: $\delta(x) = 1$ (if $x = 0$) or 0 (otherwise). Thus, the input, the desired output and the actual output of the spiking neuron are described as:

$$\begin{cases} s_i(t) = \Sigma_f \delta(t - t_i^f) \\ s_d(t) = \Sigma_g \delta(t - t_d^g) \\ s_o(t) = \Sigma_h \delta(t - t_o^h) \end{cases} \tag{4.8}$$

The products of Dirac functions are mathematically problematic. To solve this difficulty, we apply an approach called spike convolution. Unlike the method used in [15], which needs a complex error evaluation and requires spike convolution on all the spike trains of the input, the desired output and the actual output, we only convolve the input spike trains.

$$\tilde{s}_i(t) = s_i(t) * \kappa(t) \tag{4.9}$$

where $\kappa(t)$ is the convolving kernel, which we choose to be the same as Eq. (4.4). In this case, the convolved signal is in the same form as I_{PSC} in Eq. (4.3). Thus, we use I_{PSC} as the eligibility trace for the weight adaptation. The learning rule becomes:

$$\frac{dw_i(t)}{dt} = \eta[s_d(t) - s_o(t)]I_{PSC}^i(t) \tag{4.10}$$

Equation (4.10) formulates an online learning rule. The dynamics of this learning rule is illustrated in Fig. 4.2. It can be seen that the polarity of the synaptic changes depends on three cases: (1) a positive error (corresponding to a miss of the spike) where the neuron does not spike at the desired time, (2) a zero error (corresponding to a hit) where the neuron spikes at the desired time, and (3) a negative error (corresponding to a false-alarm) where the neuron spikes when it is not supposed to.

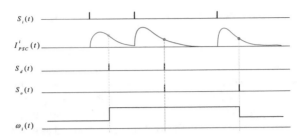

Fig. 4.2 Demonstration of the weight adaptation in PSD. $S_i(t)$ is the presynaptic spike train. $S_d(t)$ and $S_o(t)$ are the desired and the actual postsynaptic spike train, respectively. $I^i_{PSC}(t)$ is the postsynaptic current and can be referred to as the eligibility trace for the adaptation of $w_i(t)$. A positive error, where the neuron does not spike at the desired time, causes synaptic potentiation. A negative error, where the neuron spikes when it is not supposed to, results in synaptic depression. The amount of adaptation is proportional to the postsynaptic current. There will be no modification when the actual output spike fires exactly at the desired time

Thus, the weight adaptation is triggered by the error between the desired and the actual output spikes, with positive errors causing long-term potentiation and negative errors causing long-term depression. No synaptic change will occur if the actual output spike fires at the desired time. The amount of synaptic changes is determined by the current $I^i_{PSC}(t)$.

With the PSD learning rule, each of the variables involved has its own physical meaning. Moreover, the weight adaptation only depends on the current states. This is different from rules involving STDP, where both the pre- and post-synaptic spiking times are stored and used for adaptation.

By integrating Eq. (4.10), we get:

$$\Delta w_i = \eta \int_0^\infty [s_d(t) - s_o(t)] I^i_{PSC}(t) dt \qquad (4.11)$$

$$= \eta \left[\sum_g \sum_f K(t^g_d - t^f_i) H(t^g_d - t^f_i) - \sum_h \sum_f K(t^h_o - t^f_i) H(t^h_o - t^f_i) \right]$$

This equation could be used for trial learning where the weight modification is performed at the end of the pattern presentation.

In order to measure the distance between two spike trains, we use the van Rossum metric [19] but with a different filter function as described in Eq. (4.4). This filter is used to compensate for the discontinuity of the original filter function. The distance can be written as:

$$Dist = \frac{1}{\tau} \int_0^\infty [f(t) - g(t)]^2 dt \qquad (4.12)$$

where τ is a free parameter (we set $\tau = 10$ ms here), $f(t)$ and $g(t)$ are filtered signals of the two spike trains that are considered for distance measurement.

Noteworthily, this distance parameter $Dist$ is not involved in the PSD learning rule, but is used for measuring and analyzing the performance of the learning rule, which reflects the dissimilarity between the desired and the actual spike trains. In the following experiments, different values of $Dist$ are used for analysis depending on the problems. For single-spike and multi-spike target trains, we set $Dist$ to be 0.2 and 0.5, respectively, corresponding to an average time difference of around 2.5 ms for each pair of the actual and desired spikes. Smaller $Dist$ can be used if exact association is the main focus, e.g., $Dist = 0.06$ corresponds to a time difference about 0.6 ms, where no obvious dissimilarity can be seen between the two spike trains.

4.3 Results

In this section, several experiments are presented to demonstrate the characteristics of the PSD rule. The basic concepts of the PSD rule are first examined, by demonstrating its ability to associate a spatiotemporal spike pattern with a target spike train. Furthermore, we show that the PSD has desirable properties, such as generality to different neuron models, robustness against noise and learning capacity. The effects of the parameters on the learning are also investigated. Then, the application of the proposed algorithm to the classification of spike patterns is also shown.

4.3.1 Association of Single-Spike and Multi-spike Patterns

This experiment is devised to demonstrate the ability of the proposed PSD rule for learning a spatiotemporal spike pattern. The neuron is trained to reproduce spikes that fire at the same spiking time of a target train.

4.3.1.1 Experiment Setup

The neuron is connected with n afferent neurons, and each fires a single spike within the time interval of $(0, T)$. Each spike is randomly generated with a uniform distribution. We set $n = 1000$, $T = 200$ ms here. To avoid a single synapse dominating the firing of the neuron, we limit the weight below $w_{max} = 6$ nA. The initial synaptic weights are drawn randomly from a normal distribution with mean value of 0.5 nA and a standard deviation of 0.2 nA. For the learning parameters, we set $\eta = 0.01 w_{max}$ and $\tau_s = 10$ ms. The target spike train can be randomly generated, but for simplicity, we specify it as [40, 80, 120, 160] ms to evenly distribute the spikes over the whole time interval T.

4.3.1.2 Learning Process

Figure 4.3 illustrates a typical run of the learning. Initially, the neuron is observed to fire at any arbitrary time and with a firing rate different from the target train, resulting in a large distance value. The actual output spike train is quite different from the target train at the beginning. During the learning process, the neuron gradually learns to produce spikes at the target time, and that is also reflected by the decreasing distance. After finishing the first 10 epochs of learning, both the firing rate and the firing time of the output spikes match those in the target spike train. The dynamics of neuron's membrane potential is also shown in Fig. 4.3. Whenever the membrane potential exceeds the threshold, a spike is emitted and the potential is kept at reset level for a refractory period. The detailed mathematical description governing this behavior was presented previously in the section on the Spiking Neuron Model.

This experiment shows the feasibility of the PSD rule to train the neuron to reproduce a desired spike train. After several learning epochs, the neuron can successfully spike at the target time. In other words, the proposed rule is able to train the neuron to associate the input spatiotemporal pattern with a desired output spike train within

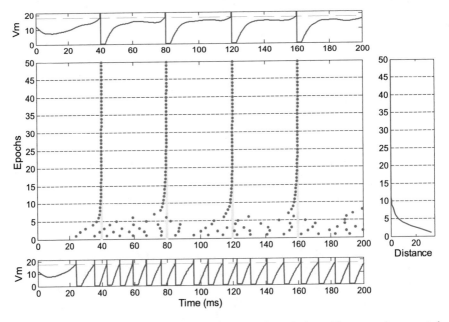

Fig. 4.3 Illustration of the temporal sequence learning of a typical run. The neuron is connected with $n = 1000$ synapses, and is trained to reproduce spikes at the target time (denoted as *light blue* bars in the *middle*). The *bottom* and *top* show the dynamics of the neuron's potential before and after learning, respectively. The *dashed red lines* denote the firing threshold. In the *middle*, each spike is denoted as a *dot*. The *right* figure shows the spike distance between the actual output spike train and the target spike train

several training epochs. The information of the input pattern is stored by a specified spike train.

4.3.1.3 Causal Weight Distribution

We further examine how the PSD rule drives the synaptic weights and the evolution of the distance between the actual and the target spike trains. In order to guarantee statistical significance, the task described in Fig. 4.3 is repeated 100 times. Each time is referred to as one run. At the initial point of each run, different random weights are used for training.

As can be seen from Fig. 4.4, the initial weights are normally distributed around 0.5 nA, which reflects the fact that there are no significant differences among the input synapses. This initial distribution of weights is expected due to the experimental setup. After learning, a causal connectivity is established. According to the learning rule, the synapses that fire temporally close to the time of the target spikes are potentiated. Those synapses that result in undesired output spikes are depressed. This temporal causality is clearly reflected on the distribution of weights after learning (Fig. 4.4). Among those causal synapses, the one with a closer spiking time to the desired time normally has a relatively higher synaptic strength. The synapses firing

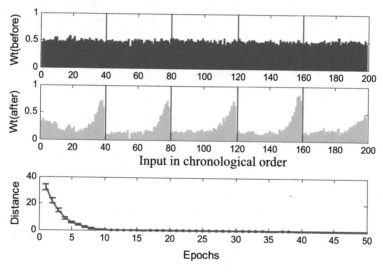

Fig. 4.4 Effect of the learning on synaptic weights and the evolution of distance along the learning process. The *top* and the *middle* show the averaged weights before and after learning, respectively. The height of each bar in the figure reflects the corresponding synaptic strength. All the afferent neurons are chronologically sorted according to their spike time. The target spikes are overlayed on the weights figure according to their time, and are denoted as *red lines*. The *bottom* shows the averaged distance between the actual spike train and the desired spike train along the learning process. All the data are averaged over 100 runs

far from the desired time will have lower causal effects. Additionally, the evolution of distance along the learning shows that the PSD rule successfully trains the neuron to reproduce the desired spikes in around ten epochs. The results also validate the efficiency of the PSD learning rule in accomplishing the single association task.

4.3.1.4 Adaptive Learning Performance

At the beginning, the neuron is trained to learn a target train as in the previous tasks. After one successful learning, the target spike train is changed to another arbitrarily generated train, where the precise spike time and the firing rate are different from the previous target. We discover that, with the PSD learning rule, we successfully train the neuron to learn the new target within several epochs. As shown in Fig. 4.5, during learning, the neuron gradually adapts its firing status from the old target to the new target.

4.3.1.5 Learning Multiple Spikes

In the scenario considered above, all afferent neurons are supposed to fire only once during the entire time window. The applicability of the PSD rule is not limited to this single spike code. We further illustrate the case where each synaptic input transmits multiple spikes during the time window. We again use the same setup as above,

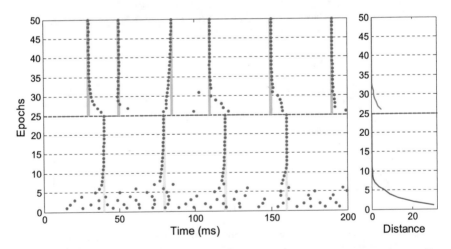

Fig. 4.5 Illustration of the adaptive learning of the changed target trains. Each *dot* denotes a spike. At the beginning, the neuron is trained to learn one target (denoted by the *light blue* bars). After 25 epochs of learning (the *dashed red line*), the target is changed to another randomly generated train (denoted by the *green* bars). The *right* figure shows the distance between the actual output spike train and the target spike train along the learning process

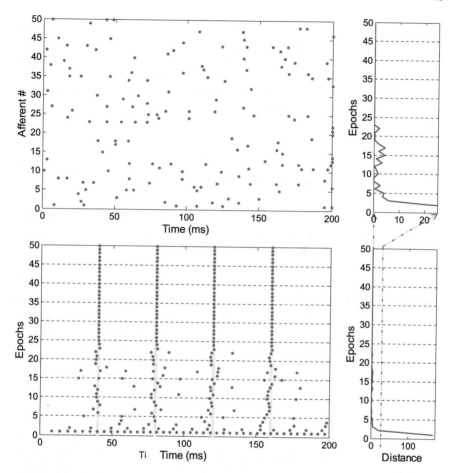

Fig. 4.6 Illustration of a typical run for learning multi-spike pattern. Each *dot* denotes a spike. The *top left* shows the input spikes from the first 50 afferent neurons out of 1000. Each synaptic input is generated by a homogeneous Poisson process with a random rate from 5–25 Hz. The *bottom left* shows the neuron's output spikes. The *right* column shows the distance between the actual output spike train and the target spike train along learning

but each synaptic input is now generated by a homogeneous Poisson process with a random rate ranging from 5–25 Hz. Multiple spikes increase the difficulty of the learning since these spikes interfere with the local learning processes [17].

As shown in Fig. 4.6, the learning although slower, is again successful. The interference of local learning processes results in fluctuations of the output spikes around the target time. In the subsequent learning epochs, the neuron gradually converges to spiking at the target time. This experiment demonstrates that the PSD rule deals with multiple spikes quite well. Compared to multiple spikes, the single spike code is simple for analysis and efficient for computation. Thus, for simplicity, we use the single spike code in the following experiments where each afferent neuron fires only once during the time window.

The above experiments clearly demonstrate that the PSD rule is capable of training the neuron to fire at the desired time. The causal connectivity is established after learning with this rule. In the following sections, some more challenging learning scenarios are taken into consideration to further investigate the properties of the PSD rule.

4.3.2 Generality to Different Neuron Models

We carry out this experiment to demonstrate that the PSD learning rule is independent of the neuron model. In this experiment, we only compare the results of learning association for the LIF and IM neuron models that were described previously. For a fair comparison, both neurons are connected to the same afferent neurons, and they are trained to reproduce the same target spike train. The setup for generating the input spatiotemporal patterns is the same as the experiment in Fig. 4.5. The connection setup is illustrated in Fig. 4.7. Except for the neuron dynamics described in Eqs. (4.1) and (4.5) respectively, all the other parameters are the same for the two neurons.

The dynamic difference between the two types of spiking neuron models is clearly demonstrated in Fig. 4.7. Although the neuron models are different, both of the neurons can be trained to successfully reproduce the target spike train with the proposed

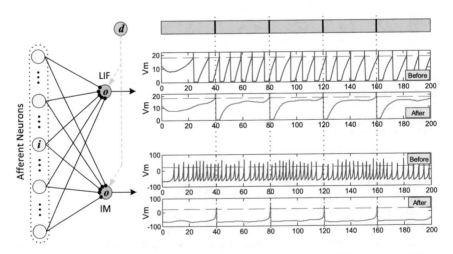

Fig. 4.7 Learning with different spiking neuron models. The LIF and IM neuron models are considered. The *left* panel shows the connection setup of the experiment. Both the two neurons are connected to the same $n = 1000$ afferent neurons, and are trained to reproduce target spikes (denoted by the *yellow* parts). The *right* panel shows the dynamics of neurons' potential before and after learning. The *dashed red lines* denote the firing threshold

PSD learning rule. It is seen that the two neurons fire at arbitrary time before learning, while after learning they fire spikes at the desired time.

In the PSD rule, synaptic adaptation is triggered by both the desired spikes and the actual output spikes. The amount of updating depends on the presynaptic spikes firing before the triggering spikes. That is to say, the weight adaptation of our rule is based on the correlation between the spiking time only. This suggests the PSD has the generality to work with various neuron models, a capability similar to that of the ReSuMe rule [17].

4.3.3   Robustness to Noise

In previous experiments, we only consider the simple case where the neuron is trained to learn a single pattern under noise-free condition. However, the reliability of the neuron response could be significantly affected by noise. In this experiment, two noisy cases are considered: stimuli noise and background noise.

4.3.3.1   Experiment Setup

In this experiment, a single LIF neuron with $n = 500$ afferent neurons is tested. Initially, a set of 10 spike patterns are randomly generated as in previous experiments. These 10 spike patterns are fixed as the templates. The neuron is trained for 400 epochs to associate all patterns in the training set with a desired spike train (the same train as is used before). Two training scenarios are considered in this experiment, i.e., deterministic training (in the noise-free condition) and noisy training. In the testing phase, a total number of 200 noise patterns are used. Each template is used to construct 20 testing patterns. We determine the association to be correct, if the distance between the output spike train and the desired spike train is lower than a specified level (0.5 is used here).

4.3.3.2   Input Jittering Noise

In the case of input jittering noise, a Gaussian jitter with a standard deviation (σ_{Inp}) is added to each input spike to generate the noise patterns. The strength of the jitter is controlled by the standard deviation of the Gaussian. The top row in Fig. 4.8 shows the learning performance. In the deterministic training, the neuron is trained purely with the initial templates. In the noisy training, a noise level of 3 ms is used. Different levels of noise are used in the testing phase to evaluate the generalization ability. For the deterministic training, the output stabilizes quickly and can exactly converge to the desired spike train within tens of learning epochs. However, the generalization accuracy decreases quickly with the increasing jitter strength. In the scenario of noisy training, although the training error cannot become zero, a better

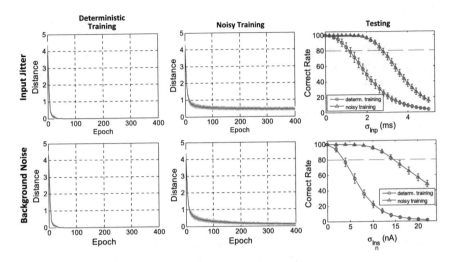

Fig. 4.8 Robustness of the learning rule against jittering noise of input stimuli and background noise. The *top row* presents the case where the noise comes from the input spike jitters. The *bottom row* presents the case of background noise. The neuron is trained under noise-free conditions (denoted as deterministic training), or is trained under noisy conditions (denoted as noisy training). In the training phase (*left two columns*), the neuron is trained for 400 epochs. Along the training process, the average distance between the actual output spike train and the desired spike train is shown. The standard deviation is denoted by the shaded area. In the testing phase (*right column*), the generalization accuracies of the trained neuron on different levels of noise patterns are presented. Both the average value and the standard deviation are shown. All the data are averaged over 100 runs

generalization ability is obtained. The neuron can successfully reproduce the desired spike train with a relatively high accuracy when the noise strength is not higher than the one used in the training. In conclusion, the neuron is less sensitive to the noise if the noisy training is performed.

4.3.3.3 Background Current Noise

In this case, the background current noise (I_{ns}) is considered as the noise source. The mean value of I_{ns} is assumed zero, and the strength of the noise is determined by its variance ($\sigma_{I_{ns}}$). A strength of 10 nA noise is used in the noisy training. We report the results in the bottom row of Fig. 4.8. Similar results are obtained as with the first case. Although the output can quickly converge to zero error in the deterministic training, the generalization performance is quite sensitive to the noise. The association accuracy drops quickly when the noise strength increases. When the neuron is trained with noise patterns, it becomes less sensitive to the noise. A relatively high accuracy can be obtained with a noise level up to 14 nA.

This experiment shows that the neuron trained under noise-free conditions will be significantly affected by noise in the testing phase. Such an influence of noise on the timing accuracy and reliability of the neuron response has been considered in many studies [5, 14, 15, 17, 20, 21]. Under the noisy training, the trained neuron demonstrates high robustness against the noise. The noisy training enables the neuron to reproduce desired spikes more reliably and precisely.

4.3.4 Learning Capacity

As used for the perceptron [22] and tempotron [5, 16] learning rules, the ratio of the number of random patterns (p) that a neuron can correctly classify over the number of its synapses (n), $\alpha = p/n$, is used to measure the memory load. An important characteristic of a neuron's capacity is the maximum load that it can learn. In this experiment, the memory capacity of the PSD rule is investigated.

4.3.4.1 Experiment Setup

We devise an experiment that has a similar setup to that in [15]. A number of p patterns are randomly generated in the same process as previous experiments, where each pattern contains n spike trains and each train has a single spike. The patterns are randomly and evenly assigned to c different categories. Here we choose $c = 4$ for this experiment. A single LIF neuron is trained to memorize all patterns correctly in a maximum number of 500 training epochs. The neuron is trained to emit a single spike at a specified time for patterns from each category. The desired spikes for the 4 generated categories are set to the time of 40, 80, 120 and 160 ms, respectively. A pattern is considered to have been correctly memorized by the neuron if the distance between the actual spike train and the desired train is below 0.2. The learning process is considered a failure if the number of training epochs reaches the maximum number.

4.3.4.2 Maximum Load Factor

Figure 4.9 shows the results of the experiment for the case of 500, 750 and 1000 afferent neurons, respectively. All the data are averaged over 100 runs. In each run, different initial weights are used. As seen from Fig. 4.9, the number of epochs required for the training increases slightly as the number of patterns increases when the load is not too high, but a sharp increase of learning epochs occurs after a certain high load. This suggests that the task becomes tougher with an increasing load. It is also noted that a larger number of synapses leads to a bigger memory capacity for the same neuron. It is reported that the maximum load factors for 500, 750 and 1000 synapses are 0.144, 0.133 and 0.124, respectively.

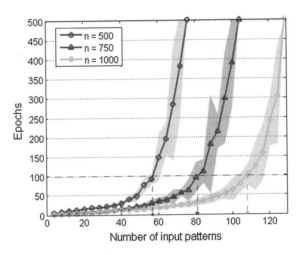

Fig. 4.9 The memory capacity of the PSD rule with different numbers of synapses. The neuron is trained to memorize all patterns correctly in a maximum number of 500 epochs. The reaching points of 500 epochs are regarded as failure of the learning. The marked lines denote average learning epochs and the shaded areas show the standard deviation. The dashed line at 100 epochs is used for evaluating the efficient load α_e described in the main text. All the data are averaged over 100 runs

4.3.4.3   Efficient Load Factor

Besides the maximum load factor, we heuristically define another factor, the efficient load α_e. The neuron can learn patterns efficiently with a relatively high load when the number of patterns does not exceed a certain value (p_e). The efficient load factor is denote as $\alpha_e = p_e/n$. When the load is below α_e, the neuron can reliably memorize all patterns with a small number of training epochs. There are different ways to define α_e. We show two possible ways. One is to derive the definition from a mathematical calculation such as $(dEpochs/dp)_{p_e} = \delta$, where δ is a specified value (for example $\delta = 0.5$). A simpler method is where a specified number of training epochs is used. The corresponding number of patterns that can be correctly learnt is considered as p_e. For simplicity, we use the latter as an example for demonstration and the specified number of epochs is set to 100. As seen from Fig. 4.9, the efficient load factors for 500, 750 and 1000 synapses are 0.112, 0.109 and 0.108, respectively. Surprisingly, these efficient load factors seem to all be around a stable value which only changes slightly across different numbers of synapses. This fixed value of efficient load factor for different values of n indicates that the number of patterns that a neuron can efficiently memorize grows linearly with the number of afferent synapses. It is worth noting that the concept of efficient load factor α_e provides an important guideline for choosing the load of patterns when a reliable and efficient training is required.

4.3.5 Effects of Learning Parameters

Two of the major parameters involved in the PSD learning rule are the learning rate η and the decay constant τ_s. In this section, we aim to investigate the effects of these parameters on the learning process.

4.3.5.1 Small τ_s Results in Strong Causal Weight Distribution

As a decay constant, τ_s is an important parameter involved in the postsynaptic current. It determines how long a presynaptic spike will still have causal effect on the postsynaptic neuron. In the phase of synaptic adaptation, τ_s also determines the magnitude of modification on the synaptic weights at the time of a triggering spike. Thus, τ_s will affect the distribution of weights after the training. To look into this effect, we conduct an experiment with a similar setup as in Fig. 4.4 but with different values of τ_s. Here we choose $\tau_s = 3$, 10 and 30 ms. As can be seen from Fig. 4.10, a smaller τ_s (3 ms) can result in a very uneven distribution with only a few synapses being given relatively higher weights. A flat distribution is obtained with an increasing τ_s. This is because τ_s determines how long the causal effect of an afferent spike will sustain. A smaller τ_s means that only the nearer neighbors are involved in generating the desired spikes, hence resulting in a smaller number of causal synapses. With a smaller number of causal synapses, a higher synaptic strength will be required to generate spikes at the desired time. On the other hand, with a larger τ_s, a wider range

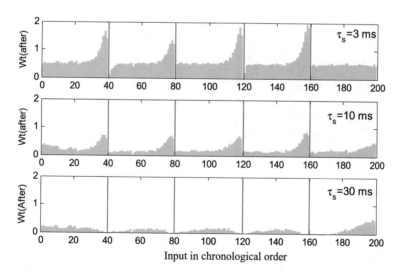

Fig. 4.10 Effect of decay constant τ_s on the distribution of weights. The averaged weights after learning are shown. The height of each bar reflects the synaptic strength. The afferent neurons are chronologically sorted according to their spike time. The target spikes are overlaid and denoted as *red lines*. Cases of $\tau_s = 3$, 10 and 30 ms are depicted. All the data are averaged over 100 runs

of causal neighbors can contribute to generating the desired spikes, and therefore a
lower synaptic strength will be sufficient. The synaptic strength and distribution for
different values of τ_s are obtained as in Fig. 4.10.

4.3.5.2 Effects of both η and τ_s on the Learning

We further conduct another experiment to evaluate the effects of both η and τ_s on
the learning. In this experiment, a single LIF neuron with $n = 500$ afferent neurons
is considered. The neuron is trained to correctly memorize a set of 10 spike patterns
randomly generated over a time window of 200 ms. The neuron is trained in a maximum number of 500 epochs to correctly associate all these patterns with a desired
spike train of [40, 80, 120, 160] ms. We denote that a pattern is correctly memorized
if the distance between the output spike train and the desired spike train is below
0.06. If the number of training epochs exceeds 500, we regard it as a failure. We
conduct an exhaustive search over a wide range of η and τ_s. Figure 4.11 shows how
η and τ_s jointly affect the learning performance, which can be used as a guidance to
select the learning parameters. With a fixed τ_s, a larger η results in a faster learning
speed (shown in Fig. 4.11, right panel), but when η is increased above a critical value
(e.g., 0.1 for $\tau_s = 30$ ms in our experiments), the learning will slow down or even
fail. For small η, a larger τ_s leads to a faster learning, however, for large η, a larger
τ_s has the opposite effect. As a consequence, when τ_s is set in a suitable range (e.g.,
[5, 15] ms), a wide range of η can result in a fast learning speed (e.g., below 100
epochs).

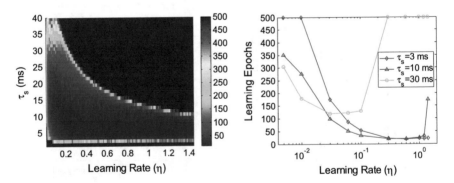

Fig. 4.11 Effects of η and τ_s on the learning. The neuron is trained in a maximum number of 500
epochs to correctly memorize a set of 10 spike patterns. The average learning epochs are recorded
for each pair of η and τ_s. The reaching points of 500 epochs are regarded as failure of the learning.
The *left* shows an exhaustive investigation of a wide range of η and τ_s, and the data are averaged
over 30 runs. A small number of learning parameters are examined in the *right* figure, and the data
are averaged over 100 runs

4.3.6 Classification of Spatiotemporal Patterns

In this experiment, the ability of the proposed PSD rule for classifying spatiotemporal patterns is investigated by using a multi-category classification task. The setup of this experiment is similar to that in [15]. Three random spike patterns representing three categories are generated in a similar fashion to that in the previous experiments, and they are fixed as the templates. A Gaussian jitter with a standard deviation of 3 ms is used to generate training and testing patterns. The training set and the testing set contain 3×25 and 3×100 samples, respectively. Three neurons are trained to classify these three categories, with each neuron representing one category. Different neurons for each category can be specified to fire different spike trains. However, for simplicity, all the neurons in this experiment are trained to fire the same spike train ([40, 80, 120, 160] ms). The experiment is repeated 100 times, with each run having different initial conditions.

After training, classification is performed on both the training and the testing set. In the classification task, we propose two decision-making criteria: absolute confidence and relative confidence. With the absolute confidence criterion, only if the distance between the desired spike train and the actual output spike train of the corresponding neuron is smaller than a specified value (0.5 is used here), then the input pattern will be regarded as being correctly classified. As for the relative confidence criterion, a scheme of competition is used. The incoming pattern will be labeled by the winning neuron that produces the closest spike train to its desired spike train.

Figure 4.12 shows the average classification accuracy for each category under the two proposed decision criteria. From the absolute confidence criterion, we see that the neuron successfully classifies the training set with an average accuracy of 99.65%.

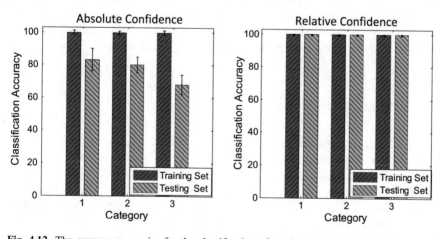

Fig. 4.12 The average accuracies for the classification of spatiotemporal patterns. There are 3 categories to be classified. The average accuracies are represented by shaded bars. Two types of criteria for making decision are proposed and investigated. The *left* is the absolute confidence criterion, and the *right* is the relative confidence criterion. All the data are averaged over 100 runs

Table 4.1 Multi-category classification of spatiotemporal patterns

Accuracy (%)	Category 1		Category 2		Category 3	
	Training	Testing	Training	Testing	Training	Testing
Absolute confidence	99.6	83.15	99.68	80.06	99.68	68.12
	±1.21	±6.79	±1.09	±4.73	±1.23	±6.09
Relative confidence	100	100	100	100	100	100
Tempotron	100	99.65	100	99.74	100	99.61
		±1.21		±1.01		±1.0

The average accuracy for the testing set is 77.11%. Noteworthily, under the relative confidence, both the average accuracies for the training and the testing set reach 100%. The performance for the classification task is therefore significantly improved by the relative confidence decision making criterion. With the absolute confidence criterion, the trained neuron strives to find a good match with the memorized patterns. However, with the relative confidence criterion, the trained neuron attempts to find the most likely category through competition.

For the classification of spatiotemporal patterns, the tempotron is an efficient rule [5] in training LIF neurons to distinguish two classes of patterns by firing one spike or by keeping quiescent. We use the tempotron rule to benchmark the PSD rule in the classification of spatiotemporal patterns. The tempotron rule is applied to perform the same classification task as above. The classification accuracies are shown in Table 4.1. As can be seen from Table 4.1, our proposed rule with the relative confidence criterion has a comparable performance to the tempotron rule. Moreover, the PSD rule is advantageous in that it is not limited to performing classification, but it is also able to memorize patterns by firing desired spikes at precise time.

4.4 Discussion and Conclusion

The PSD rule is proposed for the association and recognition of spatiotemporal spike patterns. In summary, the PSD rule transforms the input spike trains into analog signals by convolving the spikes with a kernel function. By using a kernel function, the analog signals are presented in the simple form of synaptic currents. It is biologically plausible because it allows us to interpret the signals with physical meaning. Synaptic adaptation is driven by the error between the desired and the actual output spikes, with positive errors causing LTP and negative errors causing LTD. The amount of synaptic adaptation is determined by the transformed signal of the input spikes (postsynaptic currents here) at the time of modification occurrence. When the actual spike train is the same as the desired spike train, the adaptation of the weights will be terminated.

There is a supervisory signal involved in the PSD rule. The most documented evidence for supervised rules comes from studies of the cerebellum and the cerebellar cortex [8, 9]. It is shown that supervisory signals are provided to the learning modules by sensory feedback [10] or other supervisory neural structures in the brain [9]. A neuromodulator released by the supervisory system can induce the control of the adaptation. This control occurs for several neuromodulatory pathways, such as dopamine and acetylcholine [23, 24]. Experimental evidence shows that N-methyl-D-aspartate (NMDA) receptors are critically involved in the processes of LTP and LTD [25–27]. After opening the NMDA channels, the resulting Ca^{2+} entry then activates the biochemistry of potentiation which leads to LTP [27]. Suppression of NMDA receptors by spike-mediated calcium entry may be a necessary step in the induction of LTD [27, 28]. The synaptic modification can be implemented through a supervisory control of opening or suppression of these NMDA channels.

The PSD rule is simple and efficient in synaptic adaptation. Utilizing the postsynaptic current as the eligibility trace for weight adaptation is a simple and efficient choice. The same signals of postsynaptic currents are also used in the synaptic adaptation as in the neuron dynamics, unlike the learning rules such as [12, 15, 17] where different sources of signals were used. Thus, the number of signal sources involved in the learning is reduced, which will directly benefit the computation. Secondly, unlike the arithmetic-based rules [12, 14, 15], where a complex error calculation is required for the synaptic adaptation, the PSD rule is based on a simple form of spike error between the actual and the desired spikes. The synaptic adaptation is driven by these precise spikes without complex error calculation. As a matter of fact, the weight modification only depends on currently available information (shown as Fig. 4.2). Additionally, due to the ability of the PSD rule to operate online, it is suitable for real-time applications. According to the PSD rule, different kernels, such as the exponential kernel and α kernel, can also be used in convolving the spikes to provide different eligibility traces.

The PSD rule is designed for processing spatiotemporal patterns, where the exact time of each spike is used for information transmission. The PSD rule is unsuitable for learning patterns under the rate code because this rule is designed to process precise-timing spikes by its nature. The rate code uses the spike count but not the precise time to convey information. Like other spatiotemporal mapping algorithms, including ReSuMe [13], Chronotron [14] and SPAN [15], the PSD rule cannot guarantee successful learning of an arbitrary spatiotemporal spike pattern. A sufficient number of input spikes around the desired time are required for establishing causal connections. In other words, the temporal range covered by the desired spikes should be covered by the input spikes.

In most of the experiments, a single spike code is used for afferent neurons, where each input neuron only fires a single spike during the entire time window. This single spike code is chosen for various reasons but more than one spike is also allowed for the PSD rule. Firstly, a single spike code is simple for analysis and efficient for computation. Secondly, there is strong biological evidence supporting the single spike code. The PSD rule is also suitable for multi-spike train (results shown in Fig. 4.6). When the number of spikes from each afferent neuron is not high enough, the neuron can produce the desired spike train after several epochs. When the number of spikes increases, the learning becomes slower and more difficult to converge. Additionally, the biological plausibility of an encoding scheme that can use multiple spikes to code information is still unclear.

References

1. Gerstner, W., Kistler, W.M.: Spiking Neuron Models: Single Neurons, Populations, Plasticity, 1st edn. Cambridge University Press, Cambridge (2002)
2. Ghosh-Dastidar, S., Adeli, H.: Spiking neural networks. Int. J. Neural Syst. **19**(04), 295–308 (2009)
3. Maass, W.: Networks of spiking neurons: the third generation of neural network models. Neural Netw. **10**(9), 1659–1671 (1997)
4. Shadlen, M.N., Movshon, J.A.: Synchrony unbound: review a critical evaluation of the temporal binding hypothesis. Neuron **24**, 67–77 (1999)
5. Gütig, R., Sompolinsky, H.: The tempotron: a neuron that learns spike timing-based decisions. Nat. Neurosci. **9**(3), 420–428 (2006)
6. Widrow, B., Lehr, M.: 30 years of adaptive neural networks: perceptron, madaline, and back-propagation. Proc. IEEE **78**(9), 1415–1442 (1990)
7. Knudsen, E.I.: Supervised learning in the brain. J. Neurosci. **14**(7), 3985–3997 (1994)
8. Thach, W.T.: On the specific role of the cerebellum in motor learning and cognition: clues from PET activation and lesion studies in man. Behav. Brain Sci. **19**(3), 411–431 (1996)
9. Ito, M.: Mechanisms of motor learning in the cerebellum. Brain Res. **886**(1–2), 237–245 (2000)
10. Carey, M.R., Medina, J.F., Lisberger, S.G.: Instructive signals for motor learning from visual cortical area MT. Nat. Neurosci. **8**(6), 813–819 (2005)
11. Brader, J.M., Senn, W., Fusi, S.: Learning real-world stimuli in a neural network with spike-driven synaptic dynamics. Neural Comput. **19**(11), 2881–2912 (2007)
12. Bohte, S.M., Kok, J.N., Poutré, J.A.L.: Error-backpropagation in temporally encoded networks of spiking neurons. Neurocomputing **48**(1–4), 17–37 (2002)
13. Ponulak, F.: ReSuMe-new supervised learning method for spiking neural networks. Institute of Control and Information Engineering, Poznoń University of Technology, Technical report (2005)
14. Florian, R.V.: The chronotron: a neuron that learns to fire temporally precise spike patterns. PLoS One 7(8), e40,233 (2012)
15. Mohemmed, A., Schliebs, S., Matsuda, S., Kasabov, N.: SPAN: spike pattern association neuron for learning spatio-temporal spike patterns. Int. J. Neural Syst. **22**(04), 1250,012 (2012)
16. Yu, Q., Tang, H., Tan, K.C., Li, H.: Rapid feedforward computation by temporal encoding and learning with spiking neurons. IEEE Trans. Neural Netw. Learn. Syst. **24**(10), 1539–1552 (2013)
17. Ponulak, F., Kasinski, A.: Supervised learning in spiking neural networks with resume: sequence learning, classification, and spike shifting. Neural Comput. **22**(2), 467–510 (2010)
18. Izhikevich, E.M.: Simple model of spiking neurons. IEEE Trans. Neural Netw. **14**(6), 1569–1572 (2003)
19. Rossum, M.: A novel spike distance. Neural Comput. **13**(4), 751–763 (2001)
20. Rieke, F., Warland, D., van Steveninck, R.D., Bialek, W.: Spikes: Exploring the Neural Code, 1st edn. MIT Press, Cambridge (1997)
21. Hu, J., Tang, H., Tan, K.C., Li, H., Shi, L.: A spike-timing-based integrated model for pattern recognition. Neural Comput. **25**(2), 450–472 (2013)
22. Gardner, E.: The space of interactions in neural networks models. J. Phys. **A21**, 257–270 (1988)
23. Foehring, R.C., Lorenzon, N.M.: Neuromodulation, development and synaptic plasticity. Can. J. Exp. Psychol./Rev. Canadienne de Psychologie Expérimentale **53**(1), 45–61 (1999)

24. Seamans, J.K., Yang, C.R., et al.: The principal features and mechanisms of dopamine modulation in the prefrontal cortex. Prog. Neurobiol. **74**(1), 1–57 (2004)
25. Artola, A., Bröcher, S., Singer, W.: Different voltage-dependent thresholds for inducing long-term depressiona and long-term potentiation in slices of rat visual cortex. Nature **347**, 69–72 (1990)
26. Ngezahayo, A., Schachner, M., Artola, A.: Synaptic activity modulates the induction of bidirectional synaptic changes in adult mouse hippocampus. J. Neurosci. **20**(7), 2451–2458 (2000)
27. Lisman, J., Spruston, N.: Postsynaptic depolarization requirements for LTP and LTD: a critique of spike timing-dependent plasticity. Nat. Neurosci. **8**(7), 839–841 (2005)
28. Froemke, R.C., Poo, M.M., Dan, Y.: Spike-timing-dependent synaptic plasticity depends on dendritic location. Nature **434**(7030), 221–225 (2005)

Chapter 5
A Spiking Neural Network System for Robust Sequence Recognition

Abstract This chapter presents a biologically plausible network architecture with spiking neurons for sequence recognition. This architecture is a unified and consistent system with functional parts of sensory encoding, learning and decoding. This system is the first attempt that helps to reveal the systematic neural mechanisms considering both the upstream and the downstream neurons together. The whole system is consistently combined in a temporal framework, where the precise timing of spikes is considered for information processing and cognitive computing. Experimental results show that our system can properly perform the sequence recognition task with the integration of all three functional parts. The recognition scheme is robust to noisy sensory inputs and it is also invariant to changes in the intervals between input stimuli within a certain range. The classification ability of the temporal learning rule used in our system is investigated through two benchmark tasks including an XOR task and an optical character recognition (OCR) task. Our temporal learning rule outperforms other two benchmark rules that are widely used for classification. Our results also demonstrate the computational power of spiking neurons over perceptrons for processing spatiotemporal patterns.

5.1 Introduction

As one of the cognitive abilities, sequence recognition refers to the ability to detect and recognize the temporal order of discrete elements occurring in sequence. Such sequence decoding operations are required for processing temporally complex stimuli such as speech where important information is embedded in patterns over time. However, the biophysical mechanisms by which neural circuits detect and recognize sequences of external stimuli are poorly understood.

Sequence information processing is a general problem that the brain needs to solve. Several approaches with the design of traditional artificial neural network structures [1, 2] have been considered and implemented for processing temporal information. The functionality of the brain for sequence recognition is mimicked through the artificial structures. However, these neural structures do not consider the building units of spiking neurons. Recognizing sequences of external stimuli with

© Springer International Publishing AG 2017
Q. Yu et al., *Neuromorphic Cognitive Systems*, Intelligent Systems Reference Library 126, DOI 10.1007/978-3-319-55310-8_5

spiking features in the brain still remains an open question. Numerous studies have put efforts separately to computational mechanisms with spiking neurons, where some focus on neural representations of the external information [3] while others focus on the internal procession of either upstream or downstream neurons [4–12]. Relatively few proposals exist for recognizing the sequence of incoming stimuli from a systematic level of view. Thus, a structure based on spiking neural networks is demanded. Such a spiking neural system for sequence recognition should contain several functional parts including neural coding, learning and decoding. With these functional parts integrating with each other, the system could process information from levels of upstream encoding neurons to levels of downstream decoding neurons.

Among several different temporal learning rules, without complex error calculation, the PSD rule is simple and efficient from the computational point of view, and yet biologically plausible [9]. In the classification of spatiotemporal patterns, the PSD rule can even outperform the efficient tempotron rule [9]. Moreover, the PSD rule is not limited to the classification, but can also train the neuron to associate the spatiotemporal spike patterns with the desired spike trains.

Recently, a new decoding scheme with spiking neurons has been proposed to describe how downstream neurons with dendritic bistable plateau potentials can perform the decoding of spike sequences [10, 11]. The transition dynamics of this downstream decoding network is demonstrated to be equivalent to that of a finite state machine (FSM). This decoding scheme has the same computational power as the FSM. It is capable of recognizing an arbitrary number of spike sequences [11]. However, as a part of a whole system, this decoding only describes the behavior of the downstream neurons. How the upstream neurons behave and communicate with the downstream neurons remains unclear.

In this chapter, a unified and consistent system with spiking neurons is proposed for sequence recognition. To the best of our knowledge, this is the first attempt to consider a spiking system for sequence recognition with functional parts of sensory coding, learning and decoding. This work helps to reveal the systematic neural mechanisms considering all the processes of sensory coding, learning and downstream decoding. Such a system bridges the gap between these independently studied processes. The system is integrated in a consistent scheme by processing precise-timing spikes, where temporal coding and learning are involved. The sensory coding describes how external information is converted into neural signals. Through learning, the neurons adapt their synaptic efficacies for processing the input neural signals. The decoding describes how the output neurons extract information from the neural responses. The sequence recognition of the proposed biologically plausible system is realized through the combination of item recognition and sequence order recognition. Identifying the input stimuli is required before recognizing the sequence order. The recognition scheme is robust to noisy sensory input and it is also insensitive to changes in the intervals between input stimuli within a certain range. The experiments present spiking neural networks as a paradigm which can be used for recognizing sequences of incoming stimuli.

The rest of this chapter is organized as follows. In Sect. 5.2, detailed descriptions are presented about the methods used in our integrated system. Section 5.3 shows the performances of our system through numerical simulations. Detailed investigation and analysis on different parts of the system are presented. The ability of the applied temporal learning rule is isolated for investigation using the XOR benchmark task. Then a practical optical character recognition (OCR) task is applied to investigate the functionality of our system on item recognition. The performance of the spike sequence decoding system is investigated by using a synthetic sequence of spikes. Finally, the ability of the whole system on item sequence recognition is demonstrated. Discussions about our system are presented in Sect. 5.4, followed by a conclusion in Sect. 5.5.

5.2 The Integrated Network for Sequence Recognition

In this section, the whole system for sequence recognition is described, as well as the corresponding schemes used in different subsystems.

5.2.1 Rationale of the Whole System

The whole system model contains three functional parts including sensory encoding, learning and decoding (see Fig. 5.1). They are the essential parts for a system of spiking neurons to fulfill the sequence recognition task.

In order to utilize the spiking neurons for processing external information, the first step is to get the data into them, where proper encoding methods are required [13, 14]. The components of the stimuli are connected to the encoding neurons. These encoding neurons are used to generate spatiotemporal spike patterns which represent the external stimuli. For example, each item in the sequence (denoted as '?' in Fig. 5.1) will be converted into a particular form of spike patterns by the encoding neurons. For our choice, single spikes are used, with each encoding neuron only firing once within the presence of an input item. The scheme of single spikes is simple and efficient, which would potentially facilitate computing speed since fewer spikes are involved in the computation [13, 15].

Before the recognition of the sequence order, another important step is to recognize each input item. We call this recognition process as item recognition. Without successful item recognition, it is impossible to detect the order of the incoming items since each one will be an unknown item to the system. The learning neurons in our system are used to perform the item recognition. These learning neurons are trained to perform the recognition on the encoded spike patterns that sent from the encoding neurons. During the training, the synaptic efficacies of the learning neurons are adapted to memorize the items. After training, the item recognition memory is stored in the synaptic efficacies, and thus the learning neurons can be applied in the system structure. Whenever a new incoming item comes into the system, the learning neurons can properly make a decision based on previously obtained memory.

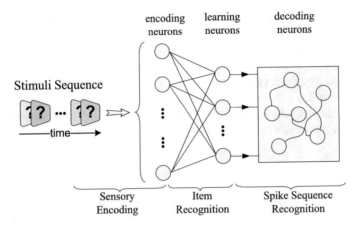

Fig. 5.1 Schematic of the system structure for sequence recognition. The system contains three functional parts which are used for sensory encoding, item recognition and spike sequence recognition, respectively. The encoding neurons convert the external stimuli into spatiotemporal spike patterns. The learning neurons would recognize the content of each input item based on the corresponding spatiotemporal spike pattern. The sequence order of the input stimuli would be recognized through the decoding neurons

After successful item recognition, the final step is to recognize the specific sequence of the incoming items. The learning neurons send spikes to the decoding neurons, where each spike represents a decision result for an input item. The target of the decoding neurons is to successfully recognize the spike sequence order of the learning neurons, and we call this recognition process as spike sequence recognition. The memory of the sequence order is stored in the connection structure of the decoding neurons.

Therefore, as is shown in Fig. 5.1, our system will process the incoming items in three main steps: (1) each item is firstly converted into spike patterns through the encoding neurons; (2) the encoded pattern will be recognized by the learning neurons, and a decision spike will be sent to the following network; (3) the desired sequence order will be recognized by the decoding neurons. In order to make the whole system function properly, the three subsystems need to communicate consistently. The functionalities of the three subsystems are encoding, item recognition and spike sequence recognition. With all these functionalities combined together, the whole system could perform the item sequence recognition. The schemes used in each subsystem are presented as follows.

5.2.2   Neural Encoding Method

Neural encoding considers how to generate a set of specific activity patterns that represent the information of external stimuli. The specific activity patterns considered

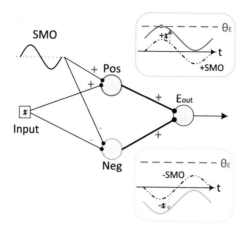

Fig. 5.2 Illustration of the phase encoding method with one encoding unit being presented. An encoding unit is composed of a positive neuron (Pos), a negative neuron (Neg) and an output neuron (E_{out}). The encoding unit receives signals from an input (x) and a subthreshold membrane potential oscillation (SMO). The value of x can be negative as well as positive. The *rectangle boxes* show the potential dynamics of the Pos and Neg neurons. The polarities of the synapses, being either positive or negative, are denoted by $+$ and $-$, respectively. Whenever the membrane potential crosses the threshold (θ_E), the neuron (Pos or Neg) will fire a spike immediately. E_{out} is strongly connected to the Pos and Neg neurons, where the firing of either the Pos neuron or the Neg neuron will immediately cause a spike from the E_{out} neuron

in this chapter are in a spatiotemporal form where precise timing of spikes is used for carrying information. Any encoding methods, that can generate spatiotemporal spike patterns with spikes being sparsely distributed within the time window, could be a proper choice for the encoding in our system. Here, we present a phase encoding method, and use it as our encoding part in the system due to its high spatial and temporal selectivity [16].

An increasing body of evidence shows that action potentials are related to the phases of the intrinsic subthreshold membrane potential oscillations ($SMOs$) [17–19]. These observations support the hypothesis of a phase code [16, 20, 21]. Such a coding method can encode and retain information with high spatial and temporal selectivity [16]. Following the coding methods presented in [16, 21], we propose a new phase encoding method. In the coding methods of [16, 21], the artificial steps of alignment and compression are questionable for biological implementation. Differently, we provide a more biologically plausible coding scheme with a clearer picture about how spikes would be generated. This would be helpful for further implementing the coding scheme on the hardware systems. Our encoding mechanism is presented in Fig. 5.2.

As can be seen from Fig. 5.2, each encoding unit contains a positive neuron (Pos), a negative neuron (Neg) and an output neuron (E_{out}). The coding units compose the encoding neurons in Fig. 5.1. Each encoding unit is connected to an input signal x and a SMO. The potentials of the Pos and Neg neurons are the summation of

x and SMO. The direction of the summation is determined by the polarities of synapses. Whenever the membrane potential firstly crosses the threshold (θ_E), the neuron will fire a spike. In order to utilize single spikes, the neuron is only allowed to fire once within the whole oscillation period T. This can be implemented through resetting the neuron's potential to prevent it from firing again within T. Due to the strong connections, the firing of either the Pos neuron or the Neg neuron will immediately cause a spike from the E_{out} neuron. The SMO for the i-th encoding unit is described as:

$$SMO_i = M \cos(\omega t + \phi_i) \tag{5.1}$$

where M is the magnitude of the SMO, $\omega = 2\pi/T$ is the phase angular velocity and ϕ_i is the initial phase. ϕ_i is defined as:

$$\phi_i = \phi_0 + (i - 1) \cdot \Delta\phi \tag{5.2}$$

where $\phi_0 = 0$ is the reference phase and $\Delta\phi$ is the phase difference between nearby encoding units. We set $\Delta\phi = 2\pi/N_{en}$ where N_{en} is the number of encoding units.

5.2.3   Item Recognition with the PSD Rule

The PSD rule [9] is recently proposed for processing spatiotemporal spike patterns. This rule is not only able to train the neurons to associate spatiotemporal spike patterns with desired spike trains, but also able to train the neurons to perform the classification of spatiotemporal patterns. As the PSD rule is simple and efficient, we use it to train the learning neurons in the proposed system for item recognition. Detailed descriptions about the PSD rule could be referred in Chap. 4.

5.2.4   The Spike Sequence Decoding Method

In this part, we describe the sequence decoding method used for the decoding neurons in our system. A network of neurons with dendritic bistable plateau potentials can be used to recognize spike sequences [11]. Based on this idea, we build our decoding system as presented in Fig. 5.3. This decoding network can recognize a specific sequence order of the spike inputs from the excitatory input neurons. The sequence scale of this network could be easily modified through adding or deleting the basic building blocks as in Fig. 5.3.

In Fig. 5.3, the dendrites have transient bistable plateau potentials, which produce the UP and DOWN states of the soma. The transitions between the UP and DOWN states are controlled by the feedforward excitation from the input neurons,

Fig. 5.3 The neural structure for spike sequence recognition. E0-5 denote the excitatory input neurons. S1-5 and D1-5 denote the soma and the dendrite respectively. Inh denotes the global inhibitory neuron. The *dashed box* shows the basic building block for scaling the network to sequences of different size

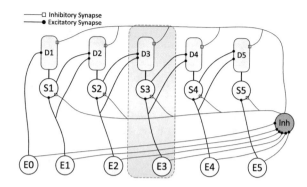

feedforward inhibition from the global inhibitory neuron, as well as the lateral excitations between the neurons in the network. There are only one global inhibitory neuron that receives input from the excitatory input neurons and sends inhibition to all the dendrite and soma. The inhibitory neuron will send a spike with a short delay after receiving an excitatory input. At the beginning of the dendrites entering the plateau potentials, they are transiently stable for a time and thus resistant to the inhibition during this time. In order to make a soma fire, there are two necessary conditions: (1) the soma is in the UP state; (2) the corresponding excitatory neuron fires. The first condition reflects the correct sequential inputs from the past and the second condition requires that the current input must be the desired one in the sequence. More detailed descriptions about the dynamics of Fig. 5.3 can be seen in Sect. 5.3.3.

The dynamics of the membrane potential of the soma is described as:

$$\tau_{sm} \frac{dV_{sm}}{dt} = -(V_{sm} - E_r) + g_{ds}(V_{dr} - V_{sm}) \qquad (5.3)$$
$$+ I_s + I_A + I_{ns}$$

where V_{sm} and V_{dr} denote the potential of the soma and the dendrite respectively; $\tau_{sm} = 20$ ms is the membrane time constant; $E_r = -70$ mV is the resting membrane potential; $g_{ds} = 0.35$ is the conductance from the dendrite to the soma; I_s is the synaptic current on the soma; I_A is the A-type potassium current; I_{ns} is a background current, and is set to zero here.

The A-type potassium current [22, 23] is activated near the resting potential and inactivated at more depolarized potentials. I_A in the soma is given by:

$$I_A = -g_A \cdot a_\infty \cdot V_{sm}^3 \cdot b(t) \cdot (V_{sm} - E_K) \qquad (5.4)$$

where $g_A = 10$ is the conductance; $E_K = -90$ mV is the reversal potential of the potassium current; a_∞ and $b(t)$ are the activation and inactivation variables respectively, and they are given by:

$$a_\infty = \frac{1}{1 + exp\left(-(V_{sm} + 70)/5\right)} \tag{5.5}$$

$$\tau_A \frac{db}{dt} = -b + \frac{1}{1 + exp\left((V_{sm} + 80)/6\right)} \tag{5.6}$$

where $\tau_A = 5$ ms is a time constant.

The synaptic current on the soma is given by:

$$I_s = -g_{As} \cdot (V_{sm} - E_E) - g_{Gs} \cdot (V_{sm} - E_I) \tag{5.7}$$

where g_{As} and g_{Gs} are the alpha-amino-3-hydroxy-5-methyl-4-isoxazolepropionic acid (AMPA) and gamma-amino-butyric-acid (GABA) synaptic conductances respectively. The AMPA and GABA synaptic conductances mediate synaptic excitation and inhibition respectively. $E_E = 0$ mV and $E_I = -75$ mV are the reversal potential of excitatory and inhibitory synapses respectively.

The dynamics of the membrane potential of the dendrite is described as:

$$\tau_{dr} \frac{dV_{dr}}{dt} = -(V_{dr} - E_r) + g_{sd} \cdot (V_{sm} - V_{dr}) + I_{dr} \tag{5.8}$$

where $\tau_{dr} = 10$ ms is the time constant of the dendrite; $g_{sd} = 0.05$ is the conductance from the soma to the dendrite; I_{dr} is the synaptic current on the dendrite, and is given by:

$$I_{dr} = - g_{Ad} \cdot (V_{dr} - E_E) - g_{Gd} \cdot (V_{dr} - E_I) \tag{5.9}$$
$$- \frac{g_{Nd} \cdot V_{dr}}{1 + exp\left(-(V_{dr} + 30)/5\right)}$$

where g_{Ad} and g_{Gd} are the AMPA and GABA synaptic conductances respectively; g_{Nd} is the N-methyl-D-aspartate (NMDA) synaptic conductance that is responsible for the transient bistable plateau potential.

An incoming spike arrives at a synapse with strength G will cause changes on synaptic conductances g: $g \rightarrow g + G$. On the dendrite, a spike to an excitatory synapse will cause $g_{Ad} \rightarrow g_{Ad} + G$ and $g_{Nd} \rightarrow g_{Nd} + 5G$. Without incoming spikes, all the synaptic conductances will decay exponentially. The decay time constants for both the AMPA and GABA conductances are 5 ms. For the NMDA conductance, the decay time constant is 150 ms. g_{Nd} is not allowed to exceed 10 due to a saturation.

The inhibitory neuron could be modeled as a single compartment quadratic LIF neuron [10, 11] such that it can respond with a short latency to an excitatory spike input. Different from this complex model, we simplify the model by setting the inhibitory neuron to spike once with a delay of 2 ms at each input spike. This choice makes the model simple and the functionality of the global feedforward inhibition remains the same.

5.3 Experimental Results

In this section, several experiments are presented to demonstrate the characteristics of our model. Through simulations, we investigate the abilities of our system mainly for item recognition and sequence recognition. A correct recognition on the input items is an essential step for further recognizing the sequence.

Firstly, in Sect. 5.3.1, the exclusive OR (XOR) problem is used to preliminarily analyze the classification ability of the temporal learning rule on spike patterns. Through the XOR task, we want to isolate the PSD rule for testing before applying it in our system. In Sect. 5.3.2, we present the ability of our system for item recognition. A set of optical characters with images of digits 0–9 are used. Section 5.3.3 shows the performance of our spike sequence decoding subsystem where the downstream neurons could recognize a specific spike sequence. Finally, in Sect. 5.3.4, the functionalities of both the item recognition and the spike sequence recognition are combined together for the item sequence recognition, which shows the performance of the whole system.

5.3.1 Learning Performance Analysis of the PSD Rule

In this part, we isolate the PSD rule from the whole system for testing with the XOR task. The XOR problem is a linearly nonseparable task, and it is a benchmark widely used for investigating the classification ability of SNNs recently [5, 24–26]. Thus, we also use the XOR problem to investigate the performance of the PSD rule firstly.

Similar to the setup in [5], we directly map the XOR inputs to spike times, with the symbol 0/1 being associated with a spike at 0/10 ms. Table 5.1 shows the input spike patterns and desired response for the XOR task. Noteworthily, the two patterns (0–0 ms and 10–10 ms) are actually identical without considering the delay of 10 ms. In order to have causal response to both inputs, a reasonable choice for the output spike should be later than 10 ms. Therefore, we use the PSD to train the neuron to fire at 12 and 22 ms for pattern (0, 0) and (1, 1), respectively. The neuron is also trained to be silent for patterns of (0, 1) and (1, 0). The PSD learning parameters are set as $\eta = 0.01$ and $\tau_s = 10$ ms.

Table 5.1 The XOR problem description

XOR input	Encoded spike input (ms)	Desired response
(0, 0)	(0, 0)	Fire
(0, 1)	(0, 10)	Silence
(1, 0)	(10, 0)	Silence
(1, 1)	(10, 10)	Fire

Fig. 5.4 The performance of the PSD rule on the XOR task with direct spike mapping. The *top panel* shows the membrane potentials of the neuron after training. The *bottom* shows the spike time of the neuron along the training. The *shaded bars* denote the desired spike time

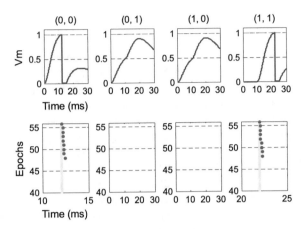

Figure 5.4 shows the PSD rule can train the neuron to solve the XOR task successfully. When patterns of (0, 1) and (1, 0) present, the neuron keeps silent without firing. For patterns of (0, 0) and (1, 1), the neuron will fire a spike as desired. However, it would be problematic if we want the neuron to fire a same desired spike train for patterns in the same group. This is due to the low dimensionality of the XOR task. Both the temporal and spatial dimensions are very limited (2 here). In order to train the neuron to fire multiple precise-timing spikes, the dimensionality of the problem should be enhanced.

Similar to the setup in [6, 26], we randomly generate two homogeneous poisson spike trains with a firing rate of 50 Hz in a time window of 200 ms. These two spike trains represent 0 or 1 respectively, and they are used to form the four inputs of the XOR problem: (0, 0), (0, 1), (1, 0) and (1, 1) (see Fig. 5.5a). We also employ the concept of reservoir computing with a network of Liquid State Machine (LSM) like in [6, 26, 27]. The LSM uses spiking neurons connected by dynamic synapses to project the inputs to a higher-dimensional feature space, which can facilitate the classification. The network used in this experiment consists of two input neurons, a noise-free reservoir with 500 LIF neurons and one readout neuron.

The target spike train could be randomly generated over the time window. For simplicity, we specify the target spike train for each category. For inputs of (0, 0) and (1, 1), the output neuron is trained to spike at [110, 190] ms, while for (0, 1) and (1, 0), it is trained to fire another target train of [70, 150] ms. Other choices of the target spikes could also be acceptable. The initial synaptic weights of the output neuron are randomly drawn from a normal distribution with a mean value of $0.5\,nA$ and a standard deviation of $0.2\,nA$. This initial condition of synaptic weights is also used for other experiments in this chapter. These synaptic weights are adjusted by the PSD rule with a set of learning parameters $\eta = 0.01$ and $\tau_s = 10$ ms. The results are averaged over 100 runs.

Figure 5.5b shows the results of a typical run, with the actual output spikes for each of the four input patterns during the learning. At the beginning, both the firing rates

Fig. 5.5 The performance of the PSD rule on the XOR task with the LSM. **a** is a general illustration of the four inputs of the XOR task. The values of 0 and 1 are represented by different spike trains. **b** shows the output spike signals for each of the four input patterns during learning in a typical run. '×' denotes the desired spike time. **c** and **d** are the results of the output neuron after 100 runs. **c** is the average spike distance between the desired and the actual output spike trains. The average spike distance for each input pattern is presented. **d** is the spike histogram showing the distribution of the actual output spikes

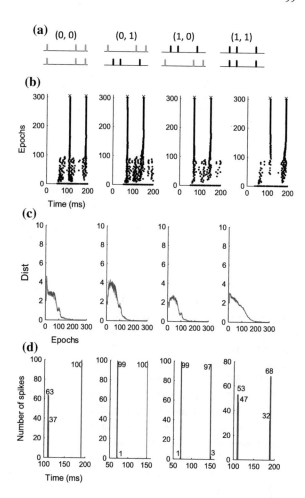

and the precise timings of the output spike trains are different from those of the target spike trains. After tens of learning epochs, the readout neuron can gradually learn to fire the target spike trains according to different input patterns. After hundreds of learning epochs, the readout neuron stabilizes at the target spike trains. This phenomenon can be also seen from the spike distance between the actual and the target spike trains (see Fig. 5.5c). A larger spike distance occurs at the beginning due to the initial conditions, followed by a gradually decreasing spike distance along the learning, and it finally converges to zero. Figure 5.5d shows the distribution of the actual output spikes corresponding to the four input patterns. From these histograms, we can see our approach with the PSD rule obtains better performance than that in [26]. Firstly, there are no undesired extra spikes or missing desired spikes in our approach. In the 100 runs of experiments, the trained neuron fires exactly 100 spikes around each desired time. Secondly, the actual output spikes are precisely and reliably

Fig. 5.6 The convergent performance. **a** shows the average spike distance over all the four input patterns. **b** is the Euclidean distance between the weights before and after each learning epoch. All the results are averaged over 100 runs

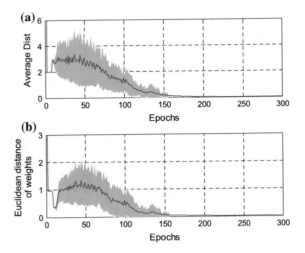

close to the desired time. The maximum error of spike time is around 1 ms. Thus, the learning success rate of our approach is higher than that in [26].

Figure 5.6 shows the convergent performance during the learning process. The average spike distance over all four input patterns is presented as well as the Euclidean distance between the weights before and after each learning epoch. As can be seen from Fig. 5.6, irregular distances occur at the first several learning epochs because of the random initial conditions. After that, the distances gradually decrease and converge to zero. The zero spike distance corresponds to the readout neuron firing exactly the target spike train, and the zero weight distance implies that there are no more changes occurring on the weights. These two distance graphs also show the ability of the PSD rule to modify the weights in order to produce the desired output spikes. Either of these two types of distance can be used as a stopping criterion for the learning process.

This experiment with the XOR problem demonstrates the ability of the PSD rule for classifying spatiotemporal patterns. The PSD rule can perform the task as desired, as long as enough dimensionality is provided by the input patterns. Considering the real-world stimuli such as images, normally, the dimensionality of the input space is not an issue. In the following, we apply the PSD rule in our system, and investigate the performance of our system for item recognition and sequence order recognition.

5.3.2   Item Recognition

In this section, the functionality of our system for the item recognition is considered. The encoding neurons and learning neurons in Fig. 5.1 are involved for the item recognition. A set of optical characters of 0–9 are used. Each image has a size of 20×20 black/white pixels, and each would be corrupted by a reversal noise where each pixel is randomly reversed with a probability denoted as the noise level. Some clean and noisy samples are demonstrated in Fig. 5.7a, b.

The phase encoding method illustrated in Fig. 5.2 is used to convert the images into spatiotemporal spike patterns. Each pixel acts as an input x to each encoding unit. x is normalized into the range of $[-0.5\theta_E, 0.5\theta_E]$ where $\theta_E = 1$ is the threshold of the encoding neurons. We set $M = 0.5\theta_E$, thus the encoded spikes will only occur at peaks of the $SMOs$. The number of encoding units N_{en} is equal to the number of pixels which is 400 here. We set the oscillation period T of the $SMOs$ to be 200 ms. The chosen scale of hundreds of milliseconds matches with the biological experimental results [3, 28]. Figure 5.7c demonstrates an encoding result with our phase coding method. The output spikes are sparsely distributed over the time window. This sparseness is compatible with the biology [29]. In addition, the sparseness could also benefit the PSD rule for constructing causal connections. However, the sparse spikes will not be obtained if we use the rank-order coding [30, 31] for the given task. All the white or black pixels will result in highly synchronized spikes. With the advantage of an additional phase coding dimension, the encoded spikes are sparsely distributed by our coding method.

We select 10 learning neurons trained by the PSD rule, with each learning neuron corresponding to one category. The learning parameters in the PSD rule are set to be $\eta = 0.06$ and $\tau_s = 10$ ms. All the learning neurons are trained to fire a target spike train with the corresponding category. The target spike train is set to be evenly distributed over the time window T (200 ms here) with a specified number of spikes n. The

Fig. 5.7 Illustration of the OCR samples. **a** shows the template images. **b** shows some image samples with different levels of reversal noise. **c** demonstrates the phase encoding result of a given image sample. Each dot denotes a spike

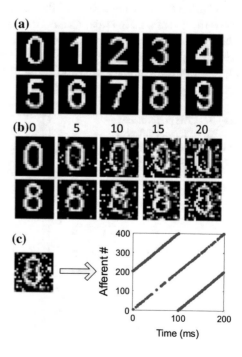

firing time of the i-th target spike is expressed as: $t_i = i/(n+1) \cdot T, i = 1, 2, \ldots n$. We choose $n = 4$ by default, otherwise will be stated. In the item recognition, the relative confidence criterion [9] is used for the PSD rule, where the incoming pattern is represented by the neuron that fires the most closest spike train to its target spike train.

In this section, several noisy scenarios are considered to evaluate the robustness of our system for item recognition: (1) spike jitter noise where a Gaussian jitter with a standard deviation (denoted as the jitter strength) is added into each encoded spike; (2) reversal noise as illustrated in Fig. 5.7b where each pixel is randomly reversed with a probability denoted as the noise level; (3) combined noise where both the jitter and the reversal noises are involved.

5.3.2.1   Spike Jitter Noise

In this scenario, the image templates are firstly encoded into spatiotemporal spike patterns. After that, jitter noises are added to generate noisy patterns. The learning neurons are trained for 100 epochs with a jitter strength of 2 ms. In each learning epoch, a training set of 100 patterns, with 10 for each category, is generated. After training, a jitter range of 0–8 ms is used to investigate the generalization ability. The number of the testing patterns for each jitter strength is set to 200. The PSD rule is applied with different numbers of target spikes ($n = 1, 2, 4, 6, 8, 10$). All the results are averaged over 100 runs.

Figure 5.8 shows the effects of the number of the target spikes on the learning performance of the PSD rule. As can be seen from Fig. 5.8, when n is low (e.g. 1, 2), the recognition performance is also relatively low. An increasing number of the target spikes can improve the recognition performance significantly (see $n = 1, 2 \rightarrow n = 4, 6$). However, a further increase in the number of target spikes ($n = 6 \rightarrow n = 8, 10$)

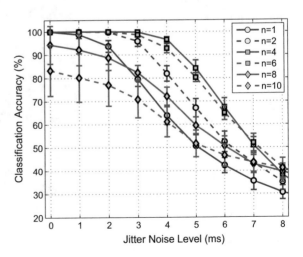

Fig. 5.8 The performance of the PSD rule with different numbers of target spikes under the case of jitter noise

Fig. 5.9 Robustness of
different rules against the
jitter noise. The PSD rule
uses $n = 4$ target spikes. The
PSD rule outperforms the
other two rules in the
considered task

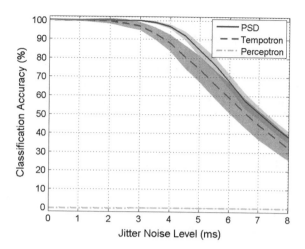

would reduce the recognition performance. The reasons for this phenomenon are due
to the local temporal features associated with each target spike. For small number of
target spikes, the neurons make decision based on a relatively less number of temporal
features. This small number of features only covers a part range of the whole time
window, which inevitably leads to a lower performance compared to a more number
of spikes. However, when the number of spikes continues increasing, an interference
of local learning processes [6] occurs and increases the difficulty of the learning.
Thus, a higher number of spikes normally cannot lead to a better performance due
to the interference. Noteworthily, compared to the idealized case of random patterns
that normally considered, the effect of the local interference is more obvious for the
learning of the practical patterns since these patterns share more common features
than random patterns.

Figure 5.9 shows the performance of different learning rules for the same classification task. We use a similar approach for the perceptron rule as in [26, 27], where the
spatiotemporal spike patterns are transformed into continuous states by a low-pass
filter. The target spike trains are separated into bins of size t_{smp}, with $t_{smp} = 2$ ms
being the sampling time. The target vectors for the perceptron contain values of 0
and 1, with 1 (or 0) corresponding to those bins that contain (or not contain) a target
spike in the bin. The input vectors for the perceptron are sampled from the continuous
states with t_{smp}. The input pattern will be classified by the winning perceptron that
has the closest output vector to the target vector.

As can be seen from Fig. 5.9, the PSD rule outperforms both the tempotron rule
and the perceptron rule. The inferior performance of the perceptron rule can be
explained. The complexity of the classification for the perceptron rule depends on
the dimensionality of the feature space and the number of input vectors for decisions.
A value of $t_{smp} = 2$ ms will generate 100 input vectors for each input pattern. These
100 points in 400-dimensional space are to be classified into 1 or 0. Learning a single
spike pattern can increase the difficulty for the perceptron rule, let alone considering

a large number of input patterns from different categories. Without separating the time window into bins, the spiking neurons by their nature are more powerful than the traditional neurons such as the perceptron. Both the PSD rule and the tempotron rule are better than the perceptron rule. The PSD rule is better than the tempotron rule since the PSD rule makes a decision based on a combination of several local temporal features over the entire time window, but the tempotron rule only makes a decision by firing one spike or not based on one local temporal feature.

5.3.2.2   Reversal Noise

In this scenario, the reversal noise is used for generating noisy patterns as illustrated in Fig. 5.7b. The learning neurons are trained for 100 epochs with a reversal noise level randomly drawn from the range of 0–10% in each learning epoch. Meanwhile, a training set of 100 noisy patterns, with 10 for each category, is generated for each learning epoch. After training, another number of 100 noisy patterns are generated and used for each reversal noise level to test the generalization ability. The noise range for testing covers 0–25% as shown in Fig. 5.10.

As can be seen from Fig. 5.10, the performances of all the three rules decrease with the increasing noise level. The performance of the PSD rule again outperforms the other two rules as in the previous scenario. Spiking neurons trained by the PSD rule can obtain a high classification accuracy (around 85%) even when the reversal noise reaches a high level (15%). The performance of the perceptron rule in this scenario is much better than that in the previous scenario. This is because of the type of the noise. The performance of the perceptron rule is quite susceptible to the changes in state vectors. Every spike of the input spatiotemporal spike patterns in the case of spike jitter noise suffers a change, while in the case of reversal noise, a change only occurs with a probability of the reversal noise level. That is to say, the elements in a

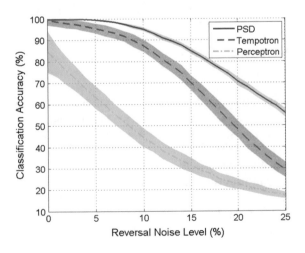

Fig. 5.10 Robustness of different rules against the reversal noise. The PSD rule uses $n = 4$ target spikes. The PSD rule outperforms the other two rules even when the noise level is high

filtered state vector have a less chance to change under the reversal noise than that under the jitter noise. Thus, the performance of the perceptron rule [26] under the reversal noise is better than that under the jitter noise.

5.3.2.3 Combined Noise

In this scenario, the jitter noise and the reversal noise are combined together to evaluate the robustness of our item recognition. Again, the learning neurons are trained for 100 epochs. In each epoch, a random reversal noise level chosen from 0 to 10% is used, as well as a jitter noise level of 2 ms. After training, a reversal noise level of 10% and a jitter noise level of 4 ms is used to investigate the generalization ability.

Figure 5.11 shows that the combined noise has a stronger impact on the performance than each single noise alone on the performance. This is expected since the effects of the two noises are combined. The perceptron rule still has a poor performance due to the jitter noise. The PSD rule still performs the best with a high average accuracy and a low deviation.

The results in this section demonstrate our item recognition with the PSD rule is robust to different noisy sensory inputs. A reliable recognition on the incoming items is essential for further sequence recognition.

5.3.3 Spike Sequence Decoding

In this section, we investigate the performance of our decoding system for spike sequence recognition. The structure of this decoding system is presented in Fig. 5.3. This decoding structure can recognize a specific sequence of E0, E1...E5. The size of the sequence can be easily extended to different scales through modification of the decoding network structure. We denote the synaptic connections as: E0→D1

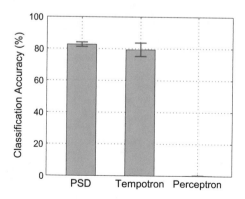

Fig. 5.11 Robustness of different rules against the combination of the jitter and reversal noises. A 10% reversal noise and a 4 ms jitter noise are used for testing

(G_0), E1-5$\rightarrow$S1-5 (G_1), E1-5$\rightarrow$D2-5 (G_2), S1-5$\rightarrow$D2-5 (G_3), Inh$\rightarrow$D1-5 (G_4), and Inh$\rightarrow$S1-5 (G_5). We set $G_0 = 5$, $G_1 = 2.5$, $G_2 = G_3 = 3$, $G_4 = 5$, $G_5 = 6$. We generate a spike input feeding into our decoding system, with Fig. 5.12 showing a 200 ms interval between nearby spikes and 230 ms for Fig. 5.13.

As can be seen from Fig. 5.12, the decoding system successfully recognizes the sequence through a firing from S5. A strong, excitatory input to the dendrite can make its potential go to a plateau potential that is transiently stable for a time. The plateau potential of the dendrite then drives the potential of the soma to a high depolarized state. Without the plateau potential of the dendrite, the potential of the soma stays near the resting potential. We refer the high depolarized state of the soma as the UP state, and the state near the resting potential as the DOWN state. Two conditions are required to make a soma fire: (1) the potential of the soma sustains in the UP state (2) when an excitatory spike input comes to this soma.

Under the experimental setup of our decoding system, the UP state of the soma can sustain for a period around 225 ms, during which the soma can reliably fire a desired spike when corresponding excitatory neuron fires. We refer this period as the reliable period. When the time interval between spikes is shorter than the reliable

Fig. 5.12 A reliable response of the spike sequence decoding system. A synthetic spike sequence is used as the input (denoted as 'Seq'). The target sequence pattern of E0, E1...E5 is highlighted by the *shaded* area. The potentials of the somas ($S1 - 5$) and the dendrites ($D1 - 5$) are shown. The interval spike time in the input sequence is 200 ms. The neurons can be successfully activated to fire when the target sequence presents

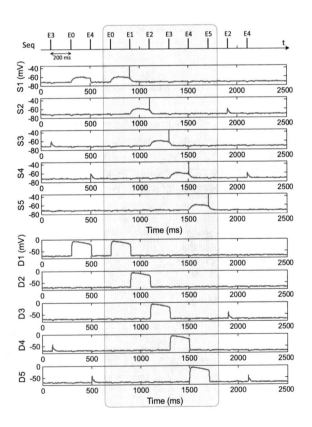

Fig. 5.13 An unreliable response of the spike sequence decoding system. The interval spike time in the input sequence is 230 ms. When the interval time is over a certain range (225 ms for this experimental setup), the neurons cannot be activated to fire even when the target sequence presents. This is because that the potential of the soma cannot sustain in the UP state for such a long interval

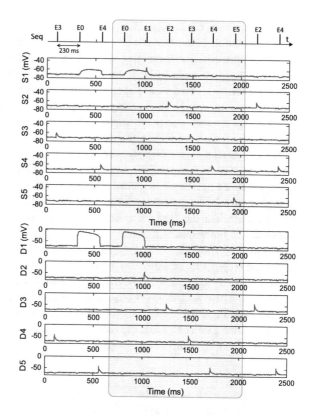

period, the decoding system can perform the recognition well (see Fig. 5.12). When the time interval between input spikes is longer than the reliable period, the UP state of a soma no longer sustains at a reliably high state. This leads to that a corresponding excitatory input spike no longer reliably drives a spike on the soma (see Fig. 5.13).

The experimental results indicate that our spike sequence decoding system is invariant to changes in the intervals between input spikes within a certain range (0–225 ms here).

5.3.4 Sequence Recognition System

In this section, the performance of the proposed whole system is investigated. The sensory encoding, temporal learning and spike sequence decoding are consistently combined together for sequence recognition. We perform the experiment with the previous digits used in Sect. 5.3.2.

These optical digits are used to form a sequence pattern, with each digit image in the sequence being corrupted by a reversal noise level of 15%. We can specify a target sequence through building connections between the output neurons of the item recognition network and the excitatory input neurons of the spike sequence decoding network. For simplicity, we specify a target sequence order of digits as: 012345. Thus, the learning neurons corresponding to the categories in this target sequence are connected to the excitatory input neurons in the sequence decoding network one by one. Each digit image is presented for 200 ms. Additionally, the interval between two successive images is not allowed to exceed 25 ms, guaranteeing a reliable performance of the spike decoding system.

We construct a sequence pattern of 6 segments, with 6 images for each segment. Every image in this sequence is randomly chosen from the 10 categories. Then the target sequence of 012345 is embedded into this sequence, with a probability of 1/3 replacing each initial segment in the sequence. After this, we feed the whole sequence to our system. The target of our system is to detect and recognize the target sequence embedded in the whole sequence.

Figure 5.14 shows the performance of our system for sequence recognition. An important step before recognizing the sequence order is to correctly recognize each input item. Only after knowing what is what, a recognition on the sequence order can be applied. The detected target sequence is represented by the firing of S5. As can be seen from Fig. 5.14, the first target sequence is successfully recognize through the sequential firing of S1-5, while the second target sequence is not correctly recognized due to a failure recognition on image '4'.

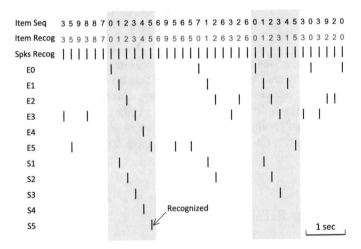

Fig. 5.14 The performance of the combined sequence recognition system. An image sequence input is fed into the sequence recognition system. Each image suffers a reversal noise of 15%. The target of this system is to detect and recognize a specified target sequence of 012345 (the *shaded* areas). 'Item Seq' denotes the input sequence of the images. 'Item Recog' is the output results of the learning neurons, with the *blue/red* color representing correct/incorrect recognition. Each output of the learning neurons results a spike in the corresponding excitatory input neurons of the spike decoding network ('Spks E'). S1-5 denote the spike output of neurons in the sequence decoding network

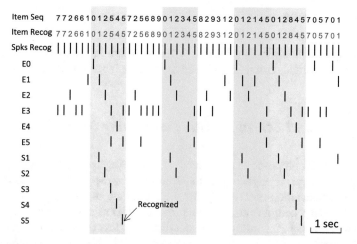

Fig. 5.15 Performance on a target sequence with one semi-blind item. The input sequence is considered in a noise-free condition. The target of this system is to detect and recognize a specified target sequence of 012?45 where '?' is from a specific range of 5–9 (illustrated in the *shaded light-cyan* areas). The *shaded light-pink* areas show some interference sequence patterns where '?' is not chosen from the allowed range

In addition, we conduct another experiment, where one item in the target sequence is semi-blind. This semi-blind item is conditioned to a specific range. We specify a target sequence of 012?45, where '?' is restricted to the range of 5–9. Other digits of '?' being out of this specific range lead to non-target sequences. For the sake of simplicity, this experiment is conducted in a noise-free condition. We randomly construct an input sequence with 48 items, and then embed the target sequences, as well as some interference sequences, into the input sequence. In order to detect and recognize the semi-blind target sequences, we reconstruct the connections between the output neurons of the item recognition network and the excitatory input neurons of the spike sequence decoding network, with all learning neurons for digit 5–9 connecting to E3 in Fig. 5.3. Other connections are not changed. As can be seen from Fig. 5.15, the semi-blind target sequences are successfully recognized, and those interference sequences are also successfully declined. Our system successfully recognizes the target sequence of 012?45 with '?' only belonging to 5–9.

These experiments show that our system with spiking neurons can perform the sequence recognition well, even under some noisy conditions. Item recognition is an essential step for a successful recognition of the target sequence. The step before recognizing the sequence order is to recognize what are the items in the input sequence. A failure recognition of the item in the target sequence would directly affect the further recognition on the sequence order.

5.4   Discussions

In this chapter, a biologically plausible system with spiking neurons is presented for sequence recognition. Discussions based on the simulation results are as follows.

5.4.1   Temporal Learning Rules and Spiking Neurons

The PSD rule [9], proposed in the concept of processing and memorizing spatiotemporal spike patterns, is applied in our system for item recognition. In the PSD rule, the synaptic adaptation is driven by the precise timing of the actual and the target spikes. Without a complex error calculation, the PSD rule is simple and beneficial for computation [9]. According to the classification tasks considered in this chapter, the PSD rule outperforms both the tempotron rule [4, 13] and the perceptron rule [26, 27].

The computational power of the spiking neurons over the traditional neurons (perceptrons) is reflected by the better performance of both the PSD rule and the tempotron rule than the perceptron rule (see Figs. 5.9 and 5.10). This is because that the spiking neurons, by their nature, are designed for processing in a time domain with a complex evolving dynamics on the membrane potential. A major difference between the perceptrons and the spiking neurons is this dynamic membrane potential. The perceptrons calculate current states in a static manner that only based on the current inputs, while the spiking neurons evolve current states in a dynamic manner that not only based on the current inputs but also the past states. Additionally, due to the ability of the spiking neurons to operate online, it can benefit the computation of a sequential procession with time elapsing.

Between the two temporal learning rules for spiking neurons, the performance of the PSD rule is better than the tempotron rule. The decisions made by the neurons under the PSD rule are based on a combination of several local temporal features over the whole time window. By contrast, the tempotron rule trains a neuron to make a decision only based on one local temporal feature if the neuron is supposed to fire a spike. A decision based on several local temporal features would result in a better performance than that only based on one local temporal feature. In addition, the PSD rule is not limited to a classification task, but it can also train a neuron to associate spatiotemporal patterns with the specified desired spike trains.

5.4.2   Spike Sequence Decoding Network

Our spike sequence decoding network is biologically realistic that can behave like FSM to recognize spike sequences [10, 11]. The functionality of this network is achieved through transitions between the UP and DOWN states of neurons. Transitions between bistable membrane potentials are widely observed through various experiments in cortical pyramidal neurons in vivo [32, 33]. The transitions between the states are controlled by feedforward excitation, lateral excitation and feedforward inhibition. The neurons enter the UP state if their dendrites have a plateau potential. The neurons will return to the DOWN state from the UP state when enough long time elapses without excitatory input spikes. In addition, the recognition is robust to time warping of the sequence. The recognition is intact as long as the interval between input spikes lies in a specific range which can be quite broad (see Figs. 5.12 and 5.13). Invariance to time warping is beneficial for tasks like speech recognition [34, 35].

5.4.3 Potential Applications in Authentication

Our system provides a general structure for sequence recognition. With proper encoding mechanisms, this system could also be applied to acoustic, tactual and olfactory signals in addition to visual signals. The processes of the item recognition and the sequence order recognition in our system could be used for user authentication to access approval. It provides a double-phase checking scheme for gaining access. Only if both the items and also their orders are correct, the person would be allowed to access.

We preliminarily applied these concepts to the speech task with our previously proposed encoding scheme [36] for sounds. The voices of ten digits were considered. It is still a very challenging task for spiking neurons to process audio signals due to variations of speed, pitch, tone and volume. Our system could be successful in the case where words are spoken in a similar manner such as samples (a)–(c) in Fig. 5.16, but it would be failed if the voice is changed significantly like (d) and (e) in Fig. 5.16. Further study is required for speech recognition with spiking neurons, and further results would be presented in our next stage.

Fig. 5.16 Voice samples of 'Zero'. **a**, **b** and **c** are samples spoken by a person in clean conditions with a similar manner for each recording. **d** is a sample under a 5dB noise and **e** is a warped sample spoken in a different manner. The *top panel* and the *bottom panel* show the sound waves and the corresponding spectrograms respectively

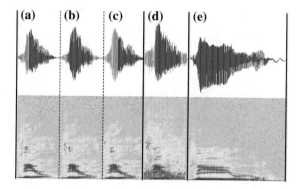

5.5   Conclusion

In this chapter, a biologically plausible network is proposed for sequence recognition. This is the first attempt to solve the sequence recognition with the network of spiking neurons by considering both the upstream and the downstream neurons together. The system is consistently integrated with functionalities of sensory encoding, learning and decoding. The system operates in a temporal framework, where the precise timing of spikes is considered for information processing and cognitive computing. The recognition performance of the system is robust to different noisy sensory inputs, and it is also invariant to changes in the intervals between input stimuli within a certain range. Our system would also be beneficial for applied developments in both hardware and software.

References

1. Starzyk, J.A., He, H.: Spatio-temporal memories for machine learning: a long-term memory organization. IEEE Trans. Neural Netw. **20**(5), 768–780 (2009)
2. Nguyen, V.A., Starzyk, J.A., Goh, W.B., Jachyra, D.: Neural network structure for spatio-temporal long-term memory. IEEE Trans. Neural Netw. Learn. Syst. **23**(6), 971–983 (2012)
3. Panzeri, S., Brunel, N., Logothetis, N.K., Kayser, C.: Sensory neural codes using multiplexed temporal scales. Trends Neurosci. **33**(3), 111–120 (2010)
4. Gütig, R., Sompolinsky, H.: The tempotron: a neuron that learns spike timing-based decisions. Nature Neurosci. **9**(3), 420–428 (2006)
5. Bohte, S.M., Kok, J.N., Poutré, J.A.L.: Error-backpropagation in temporally encoded networks of spiking neurons. Neurocomputing **48**(1–4), 17–37 (2002)
6. Ponulak, F., Kasinski, A.: Supervised learning in spiking neural networks with resume: sequence learning, classification, and spike shifting. Neural Comput. **22**(2), 467–510 (2010)
7. Florian, R.V.: The Chronotron: a neuron that learns to fire temporally precise spike patterns. PLoS One **7**(8), e40,233 (2012)
8. Mohemmed, A., Schliebs, S., Matsuda, S., Kasabov, N.: SPAN: spike pattern association neuron for learning spatio-temporal spike patterns. Int. J. Neural Syst. **22**(04), 1250,012 (2012)
9. Yu, Q., Tang, H., Tan, K.C., Li, H.: Precise-spike-driven synaptic plasticity: Learning hetero-association of spatiotemporal spike patterns. PLoS One **8**(11), e78,318 (2013)
10. Jin, D.Z.: Spiking neural network for recognizing spatiotemporal sequences of spikes. Phys. Rev. E **69**(2), 021,905 (2004)
11. Jin, D.Z.: Decoding spatiotemporal spike sequences via the finite state automata dynamics of spiking neural networks. New J. Phys. **10**(1), 015,010 (2008)
12. Byrnes, S., Burkitt, A.N., Grayden, D.B., Meffin, H.: Learning a sparse code for temporal sequences using STDP and sequence compression. Neural Comput. **23**(10), 2567–2598 (2011)
13. Yu, Q., Tang, H., Tan, K.C., Li, H.: Rapid feedforward computation by temporal encoding and learning with spiking neurons. IEEE Trans. Neural Netw. Learn. Syst. **24**(10), 1539–1552 (2013)
14. Yu, Q., Tang, H., Tan, K.C., Yu, H.: A brain-inspired spiking neural network model with temporal encoding and learning. Neurocomputing **138**, 3–13 (2014)
15. Bohte, S.M., Bohte, E.M., Poutr, H.L., Kok, J.N.: Unsupervised clustering with spiking neurons by sparse temporal coding and multi-layer RBF networks. IEEE Trans. Neural Netw. **13**, 426–435 (2002)
16. Nadasdy, Z.: Information encoding and reconstruction from the phase of action potentials. Front. Syst. Neurosci. **3**, 6 (2009)

17. Llinas, R.R., Grace, A.A., Yarom, Y.: In vitro neurons in mammalian cortical layer 4 exhibit intrinsic oscillatory activity in the 10-to 50-Hz frequency range. Proc. Natl. Acad. Sci. **88**(3), 897–901 (1991)

18. Jacobs, J., Kahana, M.J., Ekstrom, A.D., Fried, I.: Brain oscillations control timing of single-neuron activity in humans. J. Neurosci. **27**(14), 3839–3844 (2007)

19. Koepsell, K., Wang, X., Vaingankar, V., Wei, Y., Wang, Q., Rathbun, D.L., Usrey, W.M., Hirsch, J.A., Sommer, F.T.: Retinal oscillations carry visual information to cortex. Front. Syst. Neurosci. **3**, 4 (2009)

20. Kayser, C., Montemurro, M.A., Logothetis, N.K., Panzeri, S.: Spike-phase coding boosts and stabilizes information carried by spatial and temporal spike patterns. Neuron **61**(4), 597–608 (2009)

21. Hu, J., Tang, H., Tan, K.C., Li, H., Shi, L.: A spike-timing-based integrated model for pattern recognition. Neural Comput. **25**(2), 450–472 (2013)

22. Schoppa, N., Westbrook, G.: Regulation of synaptic timing in the olfactory bulb by an A-type potassium current. Nature Neurosci. **2**(12), 1106–1113 (1999)

23. Shriki, O., Hansel, D., Sompolinsky, H.: Rate models for conductance based cortical neuronal networks. Neural Comput. **15**(8), 1809–1841 (2003)

24. Ghosh-Dastidar, S., Adeli, H.: A new supervised learning algorithm for multiple spiking neural networks with application in epilepsy and seizure detection. Neural Netw. **22**(10), 1419–1431 (2009)

25. Sporea, I., Grüning, A.: Supervised learning in multilayer spiking neural networks. Neural Comput. **25**(2), 473–509 (2013)

26. Xu, Y., Zeng, X., Zhong, S.: A new supervised learning algorithm for spiking neurons. Neural Comput. **25**(6), 1472–1511 (2013)

27. Maass, W., Natschläger, T., Markram, H.: Real-time computing without stable states: a new framework for neural computation based on perturbations. Neural Comput. **14**(11), 2531–2560 (2002)

28. Butts, D.A., Weng, C., Jin, J., Yeh, C.I., Lesica, N.A., Alonso, J.M., Stanley, G.B.: Temporal precision in the neural code and the timescales of natural vision. Nature **449**(7158), 92–95 (2007)

29. Olshausen, B.A., Field, D.J.: Sparse coding with an overcomplete basis set: a strategy employed by V1? Vision Res. **37**(23), 3311–3325 (1997)

30. Van Rullen, R., Thorpe, S.J.: Rate coding versus temporal order coding: what the retinal ganglion cells tell the visual cortex. Neural Comput. **13**(6), 1255–1283 (2001)

31. Perrinet, L., Samuelides, M., Thorpe, S.J.: Coding static natural images using spiking event times: do neurons cooperate? IEEE Trans. Neural Netw. **15**(5), 1164–1175 (2004)

32. Lewis, B.L., O'Donnell, P.: Ventral tegmental area afferents to the prefrontal cortex maintain membrane potential 'up'states in pyramidal neurons via D1 dopamine receptors. Cerebral Cortex **10**(12), 1168–1175 (2000)

33. Anderson, J., Lampl, I., Reichova, I., Carandini, M., Ferster, D.: Stimulus dependence of two-state fluctuations of membrane potential in cat visual cortex. Nat. Neurosci. **3**(6), 617–621 (2000)

34. Hopfield, J.J., Brody, C.D.: What is a moment? transient synchrony as a collective mechanism for spatiotemporal integration. Proc. Natl. Acad. Sci. **98**(3), 1282–1287 (2001)

35. Gütig, R., Sompolinsky, H.: Time-warp-invariant neuronal processing. PLoS Biol. **7**(7), e1000,141 (2009)

36. Dennis, J., Yu, Q., Tang, H., Tran, H.D., Li, H.: Temporal coding of local spectrogram features for robust sound recognition. In: 2013 IEEE international conference on acoustics, speech and signal processing (ICASSP), pp. 803–807 (2013)

Chapter 6
Temporal Learning in Multilayer Spiking Neural Networks Through Construction of Causal Connections

Abstract This chapter presents a new supervised temporal learning rule for multi-layer spiking neural networks. We present and analyze the mechanisms utilized in the network for the construction of causal connections. Synaptic efficacies are finely tuned for resulting in a desired post-synaptic firing status. Both the PSD rule and the tempotron rule are extended to multiple layers, leading to new rules of multilayer PSD (MutPSD) and multilayer tempotron (MutTmptr). The algorithms are applied successfully to classic linearly non-separable benchmarks like the XOR and the Iris problems.

6.1 Introduction

In biological nervous systems, neurons communicate with others through action potentials (spikes). To emulate this phenomenon, spiking neurons are introduced to process spike information. Due to the spiking feature, the spiking neurons are more biologically plausible and computationally powerful than traditional neuron models like perceptron.

Information could be carried by spikes either in a rate-based form or a precise spike-based form. Increasing evidence shows that individual spikes with precise time play a significant role in transmitting information. Neurons can learn more and faster from the spike-based code than the rate-based code.

Considering the spatiotemporal spike patterns, many learning rules have been proposed to understand how neurons process the information. Most of temporal learning methods, such as tempotron [1], ReSuMe [2], Chronotron [3], SPAN [4] and PSD [5], only focus on the learning of single spiking neurons or single-layer SNNs. These learning rules are biologically plausible to some extent. However, the real nervous systems are extremely complex network with a large number of neurons interconnecting with each other. Investigations on the level of single neurons or single-layer networks might be insufficient to simulate the cognitive functions of the brain. Therefore, research on multilayer SNNs is demanded.

Some gradient-descent-based learning rules such as SpikeProp [6] and its extensions [7, 8] are proposed to train the network with hidden neurons to output a target

© Springer International Publishing AG 2017
Q. Yu et al., *Neuromorphic Cognitive Systems*, Intelligent Systems
Reference Library 126, DOI 10.1007/978-3-319-55310-8_6

spike train. The derivations of these rules are based on the explicit dynamics of the SRM model, which limit the applicability of these rules to other neuron models. The same problem is also involved in another gradient-descent-based rule proposed in [9]. Although the gradient-descent-based rules are effective, they lack biological explanation. The complex error calculation involved in the learning is at least questionable. In [10], an extension of the ReSuMe rule is proposed for multilayer SNNs, where the weights are updated according to STDP and anti-STDP processes. This ReSuMe-based multilayer learning rule requires back propagation of the network error. When the number of layers increases, the evaluation of the network error will become more complex. Again, such a complex error evaluation is also debatable considering the real nervous systems.

In this chapter, we propose a new supervised learning rule for multilayer spiking neural networks. This rule is an extension of the PSD rule introduced in Chap. 4. Without complex error evaluation, the learning is simple and efficient, and yet biologically plausible. In addition, we also proposed a multilayer learning for the tempotron rule. Through our multilayer learning, causal connections are constructed between layers of spiking neurons.

The rest of this chapter is organized as follows. In Sect. 6.2, the proposed learning rules for multilayer SNNs are presented, including multilayer PSD (MutPSD) rule and multilayer tempotron (MutTmptr) rule. Heuristic discussions about our multilayer learning rules are presented in Sect. 6.3. Section 6.4 presents the simulation results. Construction of the causal connections is firstly demonstrated. The properties and power of the MutPSD and MutTmptr rules are showcased by linearly non-separable tasks including the XOR problem and the Iris dataset task. Finally, discussions of the multilayer learning rules, as well as the conclusion, are presented in Sect. 6.5.

6.2 Multilayer Learning Rules

In this section, we describe the learning schemes in the feedforward multilayer spiking neural networks. Firstly, the neuron model used in this chapter is introduced. Then, the multilayer PSD (MutPSD) rule is described, followed by the introduction of multilayer tempotron (MutTmptr) rule.

6.2.1 Spiking Neuron Model

For the sake of simplicity, our neuron model consists of a leaky integrate-and-fire neuron driven by synaptic currents generated by its afferents. The potential of the neuron is a weighted sum of postsynaptic currents (PSCs) from all incoming spikes:

$$V(t) = \sum_i w_i I_{PSC}^i(t) + V_{rest} \tag{6.1}$$

where w_i and I_{PSC}^i are the synaptic efficacy and the PSC of the i-th afferent. V_{rest} is the rest potential of the neuron. The dynamics of the I_{PSC}^i is as follow:

$$I_{PSC}^i(t) = \sum_{t^j} K(t - t^j)H(t - t^j) \tag{6.2}$$

where t^j is the time of the j-th spike emitted from the i-th afferent neuron, $H(t)$ refers to the Heaviside function, K denotes a normalized kernel and we choose it as:

$$K(t - t^j) = V_0 \cdot \left(\exp(\frac{-(t - t^j)}{\tau_s}) - \exp(\frac{-(t - t^j)}{\tau_f}) \right) \tag{6.3}$$

where V_0 is a normalization factor such that the maximum value of the kernel is 1, τ_s and τ_f are the slow and fast decay constants respectively, and their ratio is fixed at $\tau_s/\tau_f = 4$.

For the neurons in the hidden layers, we utilize a fire-and-shutdown scheme as in [1]. This can guarantee a single spike scheme in the hidden neurons if the neurons receive enough strong stimulus. Increasing experimental evidence suggests that neural systems use exact time of single spikes to transmit information [11–13]. Visual system can analyze a new complex scene in less than 150 ms [11, 12]. This period of time is impressive for information processing considering billions of neurons involved. This suggests that neurons exchange only one or few spikes. In the tactile system, it is shown that the time of the first spike contains important information about the external stimuli [13]. In addition to the biological plausibility, first spikes also serve as an efficient way to transmit information. Subsequent brain region may learn more and earlier about the stimuli from the time of the first spikes [11]. The benefits of the first spike suit the role of hidden neurons acting as the information transmitter between the input and output neurons.

6.2.2 Multilayer PSD Rule

The PSD rule [5] for single neurons or single layer network is described as:

$$\frac{dw_i(t)}{dt} = \eta[s_d(t) - s_o(t)]I_{PSC}^i(t) \tag{6.4}$$

The polarity of the synaptic changes depends on three cases: (1) a positive error (corresponding to a miss of the spike) where the neuron does not spike at the desired time, (2) a zero error (corresponding to a hit) where the neuron spikes at the desired

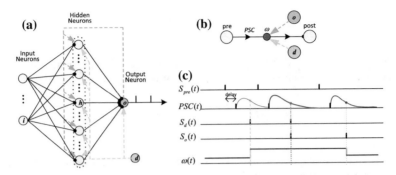

Fig. 6.1 Structure and plasticity of multilayer PSD. **a** is the structure of the multilayer network where input neurons are connected to the output neuron through hidden neurons. **b** shows the synaptic structure. The synaptic plasticity in the multilayer network is driven by the desired signal (*d*) and the actual output signal (*o*). **c** demonstrates the scheme for synaptic plasticity. A desired spike will result in potentiation, while an actual output spike will lead to depression. The amount of synaptic modification depends on the PSC signal

time, and (3) a negative error (corresponding to a false-alarm) where the neuron spikes when it is not supposed to.

In the single-layer PSD, only the direction of synaptic modification is used. The amount of modification depends on the current input PSC. Based on this idea, a multilayer PSD (MutPSD) rule can be developed. The instructor signals that only containing directions of modifications are back propagated to all synapses in the multilayer feedforward network, while the amount of synaptic change depends on the corresponding PSC received by each synapse.

Figure 6.1a shows the multilayer structure. For the reason of simplicity, one layer of hidden neurons is considered, but the algorithm can be extended to networks with more hidden layers similarly. The instructor signals are used to guide the synaptic modification direction of all synapses. Considering the synaptic delays *d*, the MutPSD rule can be described as:

$$\Delta w_i = \eta \int_0^\infty [s_d(t) - s_o(t)] I_{PSC}^i(t - d) dt \qquad (6.5)$$

$$= \eta \Big[\sum_g I_{PSC}^i(t_d^g - d) - \sum_h I_{PSC}^i(t_o^h - d) \Big]$$

where t_d^g and t_o^h denotes the *g*-th desired spike and the *h*-th actual output spike, respectively. The synaptic structure is shown in Fig. 6.1b.

The dynamics of synaptic plasticity is demonstrated in Fig. 6.1c. Similar to the single-layer PSD, the weight adaptation in the MutPSD is triggered by the error between the desired and the actual output spikes, with positive errors causing long-term potentiation (LTP) and negative errors causing long-term depression (LTD). No synaptic change will occur if the actual output spike fires at the desired time. The amount of synaptic changes is determined by the signal I_{PSC}^i.

6.2.3 Multilayer Tempotron Rule

The tempotron learning rule [1] was introduced to train a single neuron to discriminate between spatiotemporal spike patterns. Neurons are trained to distinguish between two classes by firing at least one spike or remaining quiescent. Whenever a neuron failed to fire a spike corresponding to a positive pattern, LTP will occur; if the neuron fired a spike to a negative pattern, LTD will happen.

The tempotron rule and the PSD rule are similar to some extent. In both rules, the instructor signals are used to guide the direction of the synaptic modification, either potentiation or depression. The amount of synaptic change depends on the time difference between the afferent spikes and the reference time t^{ref}. Figure 6.2 shows the learning windows. In the tempotron rule, t_{max} is the reference time for updating synaptic weights. In the PSD rule, t^{ref} refers to t_d or t_o. In the tempotron rule, it refers to t_{max}. In both the tempotron rule and the PSD rule, only the pre-synaptic spikes that precede the reference time can induce the change of synaptic weights, resulting in a construction of causal connections.

Based on the similarity with the PSD rule, a multilayer tempotron (MutTmptr) rule can be developed as an extension of the single layer tempotron. The synaptic plasticity for MutTmptr is described as follow:

$$\Delta w_i = \begin{cases} \eta \sum_{t_i < t_{max}} K(t_{max} - t_i - d), & \text{if } P^+ \text{ error;} \\ -\eta \sum_{t_i < t_{max}} K(t_{max} - t_i - d), & \text{if } P^- \text{ error;} \\ 0, & \text{otherwise.} \end{cases} \tag{6.6}$$

where t_{max} denotes the time at which the neuron reaches its maximum potential value in the time domain, and d denotes the synaptic time delay. The above equation is equivalent to the follow:

$$\Delta w_i = \begin{cases} \eta \cdot I^i_{PSC}(t_{max} - t_i - d), & \text{if } P^+ \text{ error;} \\ -\eta \cdot I^i_{PSC}(t_{max} - t_i - d), & \text{if } P^- \text{ error;} \\ 0, & \text{otherwise.} \end{cases} \tag{6.7}$$

Fig. 6.2 Similarity between the PSD rule and the tempotron rule on learning windows. The amount of synaptic change depends on the time difference Δt between the afferent spikes t_{pre} and the reference time t^{ref}

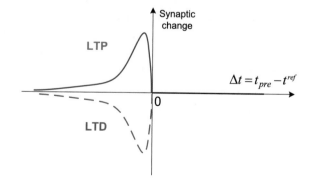

where I^i_{PSC} denotes the post-synaptic current (PSC) of the corresponding synapse.

The instructor signal, containing the modification direction, is back propagated to all synapses in the multilayer network. The amount of synaptic change depends on the PSC signal. Equations 6.5 and 6.7 are used to conduct the following simulations.

6.3 Heuristic Discussion on the Multilayer Learning Rules

In this section, we use a simple three-layer network (see Fig. 6.1) to analyze the process of the synaptic modification in our MutPSD and MutTmptr rules. For simplicity, neurons are connected through synapses without delays. Synaptic change between the input and the hidden neurons is denoted as Δw_{ih}. Δw_{ho} refers to the change between the hidden and the output neurons.

The following cases would occur along the learning.

1. The output neuron fires a spike at t_o in MutPSD or fires a spike to negative patterns in MutTmptr:
   The LTD process will occur. The depression is back propagated to all synapses, resulting in $\Delta w_{ih} < 0$ and $\Delta w_{ho} < 0$. The excitatory synapses will become less excitatory and the inhibitory synapses will become more inhibitory. This could eliminate the wrong spike of the output neuron. A decrease in w_{ho} could cause the decrease in the potential of the output neuron, thus the spike could be eliminated. The decrease in w_{ih} would result in a silent response of the hidden neuron. Without the stimulating signal (spikes) from the hidden neuron, the output neuron could become silent as desired.

2. The output neuron fails to fire a spike at t_d in MutPSD or keeps silent to a positive pattern in MutTmptr:
   The LTP process will occur. Similar to the depression process, the potentiation is back propagated to all synapses, resulting in $\Delta w_{ih} > 0$ and $\Delta w_{ho} > 0$. The excitatory synapses will become more excitatory and the inhibitory synapses will become less inhibitory. As a result, the potential of the output neuron could be increased, leading to a spike correspondingly.

3. The output neuron reacts correctly as desired:
   In the MutPSD rule, this means the output neuron only fires at the time of t_d. In the MutTmptr rule, it means the output neuron fires at least one spike to positive patterns and keeps silent to negative patterns. If the output neuron responds as desired, no synaptic modification occurs.

The instructor signal guides the direction of the synaptic modification, leading the output neuron to a desired response if such a solution exists.

6.4 Simulation Results

In this section, several simulation experiments are conducted to demonstrate the capabilities of the algorithm. Firstly, through the association of spatiotemporal spike pattern by the MutPSD rule, we demonstrate how the causal connections are constructed. Both the MutPSD and the MutTmptr rules are then applied to classic benchmarks, including the XOR problem and the Iris dataset.

6.4.1 Construction of Causal Connections

In order to demonstrate the construction of causal connections, the MutPSD rule is used to train the neuron to associate the input spatiotemporal spike pattern with a desired spike train.

6.4.1.1 Technical Details

We construct the network in the structure of $50 \times 100 \times 1$, without the synaptic delay. The input spatiotemporal spike pattern connects with the network through the input neurons. The spatiotemporal spike pattern is designed in a single-spike manner, where each input neuron only fires once within a time window of 30 ms. The output neuron is trained, within a max number of training epochs (150), to fire a desired spike train of [10, 20, 30] ms. The initial weights are uniformly distributed in the range of [0, 0.5]. We set $\eta = 0.01$ and $\tau_s = 7$ ms. The learning is considered converged when each of the actual output spike approaches to the corresponding desired spike within a precision of 0.1 ms.

6.4.1.2 Analysis of the Learning

Figure 6.3a shows the input spatiotemporal spike pattern. The network is trained to associate this spike pattern with the desired spike train. As is shown in Fig. 6.3c, the output neuron gradually learns to fire at the desired times. At the begin, both the firing rate and the precise time of the output spikes are different from those of the desired spikes. Along the learning, the output neuron can successfully fire the desired spikes. This can also be reflected through the spike distance graph, where a small distance denotes a big similarity between the desired and the actual output spike trains.

Figure 6.3b shows the firing behavior of the hidden neurons. Before learning, the spikes of the hidden neurons are far away from the desired time, thus it is difficult for these hidden spikes causing desired output spikes. After learning, a sufficient number of hidden spikes appear before each desired spike. These hidden spikes are

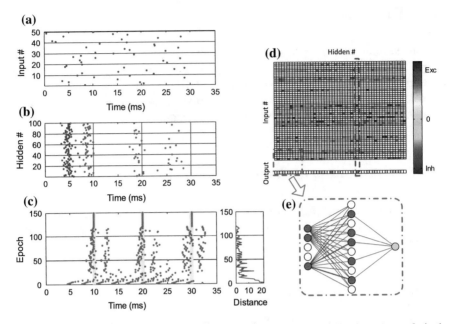

Fig. 6.3 Construction of causal connections. The multilayer network is trained to output a desired spike train associating with the input pattern. **a** is the demonstration of the input spatiotemporal spike pattern. **b** shows the spikes of the hidden neurons before learning (*blue*) and after learning (*magenta*). *Vertical red lines* denote the target time. **c** shows the actual output spikes along the learning epochs. The spike distance between the desired and actual output spike trains is also shown. *Shaded bars* denote the desired spike time. **d** demonstrates the weight matrix of the network that relating to the target time of 10 ms. The intensity reflects the weight value, and white boxes mean the corresponding connections do not fire before this target time. **e** demonstrates a connection view of the corresponding part in **d**. *Shaded* neurons mean that they fired before 10 ms, thus they are wired up to construct causal connections. The weight strength is denoted by the line width

necessary for resulting in spikes at the desired times. We denote those pre-synaptic spikes that resulting in a post-synaptic spike as the causal spikes. Another necessary factor for causing desired spikes is that the synaptic weights corresponding to the causal spikes should be fine tuned.

Figure 6.3b demonstrates one necessary factor with respect to the causal spikes. The other necessary factor regarding to the weights are shown in Fig. 6.3d, e. For simplicity, only the causal connections for firing a target spike at 10 ms are shown. Figure 6.3d shows the weight matrix of the network. For example, the red rectangle reflects the weights relating to a specific hidden neuron, with upper figure showing connections from input neurons to this hidden neuron, and lower figure representing the weight from this hidden neuron to the output neuron. In the weight matrix figure, the white boxes mean the corresponding connections do not have causal spikes. Figure 6.3e shows the connection structure corresponding to the part in Fig. 6.3d. Neurons without causal spikes do not have effect on the desired spikes. As can be seen from the figure, causal neurons are connected with fine tuned weights, including both the excitatory and inhibitory synapses.

6.4.2 The XOR Benchmark

The XOR problem is a linearly nonseparable task, and it is a classic benchmark problem widely used for investigating the classification ability of spiking neural networks recently [6, 8, 10, 14]. Thus, we also use the XOR task to investigate the ability of our MutPSD and MutTmptr rules in this section. Detailed experimental setup and results are presented as follows.

6.4.2.1 Technical Details

Similar to [6], the input and output patterns for the XOR task are encoded into spikes (as can be seen in Table 6.1). The XOR input of 0/1 is directly converted to the spike input of 0/6 ms. In addition to these two inputs, a third neuron with an input spike at 0 ms is used to serve as the time reference. Without this time reference, pattern (0, 0) and (1, 1) would be identical in the view of spikes, thus the network would be unable to distinguish them.

We choose the network structure as $3 \times 5 \times 1$. Additionally, multiple sub connections (mSub) with different delays were used. We set mSub = 15, with delays ranging from 0 to 12 ms. The network was trained with $\eta = 0.01$ and $\tau_s = 7$ ms, otherwise will stated. The network was simulated with a time window of 30 ms and a time step of 0.1 ms.

6.4.2.2 Demonstration of the Learning

The capabilities of both the MutPSD and the MutTmptr rules on the XOR task are demonstrated here. In the MutPSD rule, the output neuron is required to fire desired spikes with a precision of 0.2 ms corresponding to different input patterns. In the

Table 6.1 XOR problem description for multilayer SNNs

XOR input	Encoded spike input (ms)			MutPSD output (ms)	MutTmptr output
(0, 0)	0	0	0	16	Fire
(0, 1)	0	6	0	10	Silent
(1, 0)	6	0	0	10	Silent
(1, 1)	6	6	0	16	Fire

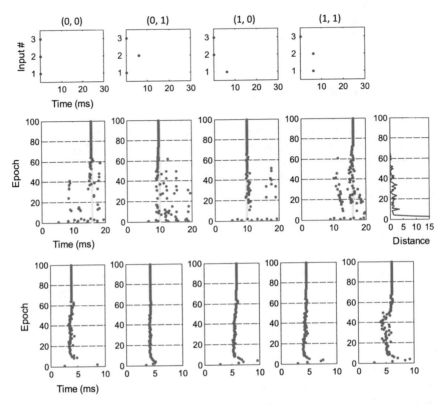

Fig. 6.4 Demonstration of the MutPSD rule for the XOR task. The *top row* shows the four spike input patterns of the XOR task. The *middle row* shows the actual output spikes according to each input pattern. The *shaded bars* denote the desired spike time. The average spike distance is also shown on the right. The *bottom row* shows the output spikes of the hidden neurons for the input pattern of (0, 0)

MutTmptr rule, instead of firing precisely, the output neuron is only required to correctly fire or not fire corresponding to an input pattern.

Figure 6.4 demonstrates the MutPSD rule can successfully train the network to learn the XOR task. As is shown in Chap. 5, the single-layer PSD cannot directly learn this task, unless a reservoir network is used to enrich the dimension of the input space. In our MutPSD rule, a small number of hidden neurons with adjustable weights are sufficient for the XOR task. Along the learning, the output neuron gradually learns to fire desired spikes corresponding to different input patterns. The spike distance between the desired and the actual output spikes decreases gradually. The synaptic efficacies of the hidden neurons are also modified along the learning, which is reflected from the adjustment of their spike times. The adjusted spike time of the hidden neurons can facilitate the output neuron to fire desired spikes. These hidden spikes serve as the stimulating sources for the output neuron.

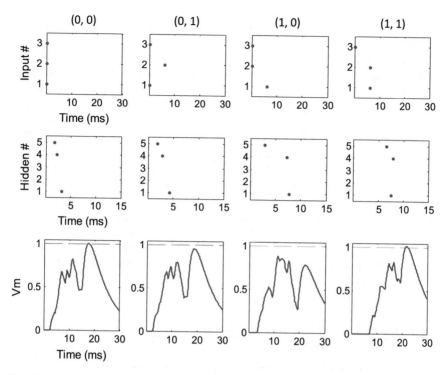

Fig. 6.5 Demonstration of the MutTmptr rule for the XOR task. The *top row* shows the four spike input patterns of the XOR task. The *middle row* shows the actual output spikes of the hidden neurons according to each input pattern. The *bottom row* shows the membrane potentials of the output neuron for each corresponding input pattern

As can be seen from Fig. 6.5, the MutTmptr rule can also train the network to perform the XOR task well, with the output neuron firing a spike for patterns of (0, 0) and (1, 1), and keeping silent for (0, 1) and (1, 0). The hidden neurons fire differently for each input pattern. Again, these spikes from the hidden neurons serve as the stimulating sources for the output neuron. Noteworthily, although 5 hidden neurons are chosen for the XOR task, only a small number of these hidden neurons (#1, #4, #5) are utilized. Therefore, our multilayer learning rule can effectively select a sufficient number of resources that are enough to fulfill the task.

6.4.2.3 Convergence of the Learning

In order to investigate the convergent performance of our multilayer learning rules, the previous demonstration experiment is conducted for 50 runs. For the MutPSD rule, a precision of 1 ms is used as in other studies [10, 15]. The average results are reported in Table 6.2.

Table 6.2 shows our multilayer learning rules are more efficient for the XOR task. To train the output neuron to spike precisely corresponding to different patterns, our MutPSD rule has a faster convergent speed compared to other rules. A higher successful rate of runs is also obtained compared to that in [10]. In addition, with less number of learning parameters, our MutPSD rule is simpler compared to multilayer ReSuMe rule in [10]. Regardless of firing precisely, the MutTmptr rule converges even faster. This is expected since only a response of fire or not fire is considered for the MutTmptr rule. Such a binary response can simplify the learning compared to those rules for precise response in time.

Figure 6.6 shows the effect of the learning rate η on our multilayer learning rules. As can be seen in this figure, a smaller η results in a slower learning speed. The learning becomes faster with an increasing η. However, a further increase in η cannot benefit the learning. The successful rate of runs can be decreased with a larger η (results are not shown here). Additionally, the learning speed of the MutTmptr rule is always faster than that of the MutPSD rule. As discussed before, this is because the MutTmptr rule only needs to train the neuron to have a binary response of fire or not, regardless of the precise time of the response.

Table 6.2 Convergent results for the XOR problem

	Precision of convergence (ms)	No. of epochs for convergence	Successful rate (%)
Bohte [6]	0.71	250	–
McKennoch [15]	1.0	127	–
Sporea [10]	1.0	137	98
MutPSD	1.0	86	100
MutTmptr	–	37	100

Fig. 6.6 Effect of the learning rate on the convergence of the XOR task. Results are averaged over 50 runs

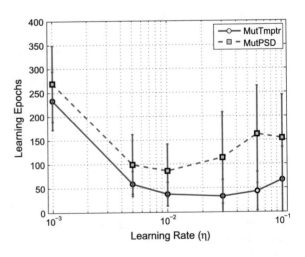

6.4.3 The Iris Benchmark

In order to investigate the recognition performance of our multilayer rules, the classic Iris benchmark task is considered. The dataset consists of three classes of Iris flowers, with one class being linearly separable from the other two classes, and two classes being nonlinearly separable with each other. Each class contains 50 samples and each sample is represented by 4 variables.

6.4.3.1 Technical Details

To encode the variables of Iris, we use the same population encoding scheme as in [6, 9], where each feature is encoded separately by an array of Gaussian functions with different centers. For a variable x in a range $[x_{min}, x_{max}]$, n neurons with different Gaussian receptive fields are used to encode. The center and width of the i-th neuron are set to $\mu_i = x_{min} + (2 \cdot i - 3)/2 \cdot (x_{max} - x_{min})/(n - 2)$ and $\sigma_i = 1/1.5 \cdot (x_{max} - x_{min})/(n - 2)$, respectively. Each feature is encoded as n (set as 5) spike times between 0 and 10 ms. Thus, the total number of input neurons is $4 \times 5 + 1 = 21$. The number of hidden neurons is selected as 8. The number of sub connections is set to 5, and each synapse has a synaptic delay between 0 and 10 ms. Three networks of $21 \times 8 \times 1$ are constructed with each network for one class. The upper limit of training epochs is set to 300. For the MutPSD rule, each network is trained to fire a desired train of [15, 25] ms corresponding to the correct input class, and to keep silent for other classes. In the MutTmptr rule, each network is trained to fire a spike for the positive class, and to keep silent for other classes.

6.4.3.2 Analysis of the Learning

We use a winner-take-all scheme for the readout. For the MutPSD rule, the network with closest spike distance dominates the class of the input pattern. For the MutTmptr rule, two different winner-take-all readout schemes are investigated. One regards to the fire status (denoted as MutTmptr_Fire), and the other one regards to the maximum potential (denoted as MutTmptr_Vmax).

As can be seen from Fig. 6.7, the MutTmptr rule can learn the training set better than the MutPSD rule, while the MutPSD rule has a better generalization performance. It can be seen from Fig. 6.7b, the testing accuracy tends to increase with the increasing number of samples used for training. If only output spikes are considered for the readout, the MutPSD rule performs better than the MutTmptr_Fire rule. This is because the MutPSD rule makes decision based on a combination of several local temporal features, but the MutTmptr rule only uses single temporal feature for the decision. In addition, the MutTmptr rule requires all the three nets to response cor-

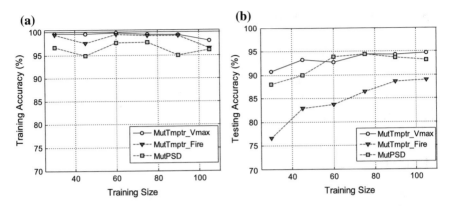

Fig. 6.7 Performance of multilayer learning rules on the Iris task. **a** and **b** show the training and testing accuracy, respectively. Results are averaged over 10 runs

rectly for a correct decision. This is another factor affects its performance. If we use the maximum potential for the decision, the performance is improved significantly (see MutTmptr_Vmax in Fig. 6.7b).

6.5   Discussion and Conclusion

In this chapter, we proposed two learning rules for multilayer SNNs, namely the multilayer PSD rule (MutPSD) and the multilayer tempotron rule (MutTmptr). These two rules are similar, where a supervisor signal, containing the synaptic modification direction, is back propagated to the synapses in the network. Without complex error evaluation as in [6, 9, 10], our multilayer learning rules are simpler and more efficient. In addition, it is not biologically plausible for the neurons to back propagated a calculated error, or it is at least questionable. A global neuromodulatory signal, determining the polarity of the synaptic changes, would be more feasible [1].

The amount of synaptic change depends on the pre-synaptic currents. This scheme, combined with the supervisor signal, can help to construct the causal connections between neurons. Correlated neurons are connected with fine tuned weights, resulting in a desired response at the output neuron. The hidden neurons serve as the information transmitter between the input and output neurons.

The MutTmptr rule has a faster convergent speed than the MutPSD rule. This is because the MutTmptr rule only trains the network to respond correctly with a binary status, either fire or not. For the MutPSD rule, the precise spike time of the output neuron is also considered. This makes the learning more difficult than the MutTmptr rule. However, the MutPSD rule has a better generalization ability compared to the MutTmptr rule. This is due to that, the MutPSD makes a decision based on a combination of several local temporal features, while the MutTmptr uses only a single local temporal feature for a decision.

In summary, both the MutPSD and the MutTmptr rules are simple, efficient and yet biologically plausible. We demonstrate the mechanisms that how the causal connections are constructed in the multilayer spiking neural networks. The performances of our multilayer learning rules are investigated through the two classic benchmark tasks, that is the XOR task and the Iris dataset problem. The MutTmptr rule can provide a faster learning speed, while the MutPSD rule gives a better generalization ability.

References

1. Gütig, R., Sompolinsky, H.: The tempotron: a neuron that learns spike timing-based decisions. Nat. Neurosci. **9**(3), 420–428 (2006)
2. Ponulak, F.: ReSuMe-new supervised learning method for spiking neural networks. Institute of Control and Information Engineering, Poznoń University of Technology, Tech. rep. (2005)
3. Florian, R.V.: The Chronotron: a neuron that learns to fire temporally precise spike patterns. PLoS One **7**(8), e40,233 (2012)
4. Mohemmed, A., Schliebs, S., Matsuda, S., Kasabov, N.: SPAN: spike pattern association neuron for learning spatio-temporal spike patterns. Int. J. Neural Syst. **22**(04), 1250,012 (2012)
5. Yu, Q., Tang, H., Tan, K.C., Li, H.: Precise-spike-driven synaptic plasticity: Learning hetero-association of spatiotemporal spike patterns. PLoS One **8**(11), e78,318 (2013)
6. Bohte, S.M., Kok, J.N., Poutré, J.A.L.: Error-backpropagation in temporally encoded networks of spiking neurons. Neurocomputing **48**(1–4), 17–37 (2002)
7. Ghosh-Dastidar, S., Adeli, H.: Improved spiking neural networks for EEG classification and epilepsy and seizure detection. Integr. Comput.-Aided Eng. **14**(3), 187–212 (2007)
8. Ghosh-Dastidar, S., Adeli, H.: A new supervised learning algorithm for multiple spiking neural networks with application in epilepsy and seizure detection. Neural Netw. **22**(10), 1419–1431 (2009)
9. Xu, Y., Zeng, X., Han, L., Yang, J.: A supervised multi-spike learning algorithm based on gradient descent for spiking neural networks. Neural Netw. **43**, 99–113 (2013)
10. Sporea, I., Grüning, A.: Supervised learning in multilayer spiking neural networks. Neural Comput. **25**(2), 473–509 (2013)
11. Gollisch, T., Meister, M.: Rapid neural coding in the retina with relative spike latencies. Science **319**(5866), 1108–1111 (2008)
12. Thorpe, S., Fize, D., Marlot, C.: Speed of processing in the human visual system. Nature **381**(6582), 520–522 (1996)
13. Johansson, R.S., Birznieks, I.: First spikes in ensembles of human tactile afferents code complex spatial fingertip events. Nat. Neurosci. **7**(2), 170–177 (2004)
14. Xu, Y., Zeng, X., Zhong, S.: A new supervised learning algorithm for spiking neurons. Neural Comput. **25**(6), 1472–1511 (2013)
15. McKennoch, S., Liu, D., Bushnell, L.G.: Fast modifications of the spikeprop algorithm. In: International Joint Conference on Neural Networks, 2006. IJCNN'06, pp. 3970–3977. IEEE (2006)

Chapter 7
A Hierarchically Organized Memory Model with Temporal Population Coding

Abstract Memory is a critical process in the brain to many cognitive behaviors. It is a complex process operating across different brain regions. However, the organizing principles of memory systems remain unclear. Emerging experiment results show that memories are represented by population of neurons and organized in a categorical and hierarchical manner. In this work, we describe a hierarchically organized memory (HOM) model using spiking neurons, in which temporal population codes are considered as the neural representation of information and spike-timing-based learning methods are employed to train the network. The results have demonstrated that memory coding units are formed into neural cliques, and information are stored in the form of associative memory within gamma cycles. Moreover, temporally separated patterns can be linked and compressed via enhanced connections among neural groups forming episodic memory. Our model provides a computational interpretation of memory organization at a system level.

7.1 Introduction

Memory is an extremely complex and brain-wide process, which is an indispensable part of what makes various intelligence. Researchers have devoted significant effort to investigating what is memory and how it works. In order to understand amazing functions of human memory, two basic questions should be answered in the first place: what is the internal representation of memory and how memory is organized in the brain?

During the last few decades, a great deal of works have been conducted to investigate how information is represented in the nervous system. As a traditional coding scheme, rate coding assumes that the most important information about a stimulus can be described by the firing rates of sensory neurons. However, rate codes fail to describe rapidly varying real-world stimuli. Recent experimental studies show that spike timing makes sense in visual [1], auditory [2], olfactory [3] pathways and hippocampus [4] in various neuronal systems [5]. It has been reported that precisely timed spikes play a pivotal role during the integration process of cortical neurons [6]. In addition, studies of population coding suggest that information can be encoded by

© Springer International Publishing AG 2017
Q. Yu et al., *Neuromorphic Cognitive Systems*, Intelligent Systems
Reference Library 126, DOI 10.1007/978-3-319-55310-8_7

clusters of cells rather than single cells. Population coding has been found existing throughout the nervous system. Visual features are encoded with population codes in the visual cortex [7], movement directions are found to be related to populations of motor cortical neurons [8], and place cells have been identified when an animal passes by a specific location in an environment [9, 10]. Recently, temporal population coding has been demonstrated to be capable of encoding visual stimuli invariantly [11]. We believe that the combination of both temporal codes and population codes provides an alternative approach to achieving neural representation of information in the nervous system.

The organization of memory is closely associated with learning process that never ceases throughout the brain. With the development of large-scale ensemble recording techniques, network-level functional coding units have been identified in the hippocampus [12]. Moreover, a recent study on population response patterns in monkey inferior temporal cortex suggests that external stimuli can be represented by responses of neural populations in monkey inferior temporal (IT) cortex, and encoded memory patterns are organized in a hierarchical order with combination of neural cliques [13, 14]. As a special form of Hebbian learning, STDP process and other spike-timing based learning schemes are believed to be involved in the formation of neural cliques and associative learning. Although different spiking neural network memory models and learning algorithms have been proposed, few of them employ temporal codes as the neural representation or pay enough attention to memory organization.

Researchers have proposed various working memory models exploring different mechanisms to achieve persistent neural activity (for review please refer to [15, 16]). Different learning algorithms for spiking neuron have been proposed to study hetero-association [17–22]. Whereas, memory models concerning auto-associative memory and episodic memory remains underexplored. Moreover, among existing memory models using spiking neural network, how to realize memory function with temporal codes and how memory is organized in nervous system still need more investigation.

The *Cortext* model [23], which is inspired by the anatomical structure of the cerebral cortex, is known as a hierarchical model for word recognition. The *Cortext* model consists of three layers mimicking V1, V2 and IT, while the information processing mechanism is based on a predictive coding scheme. In one macrocolumn, which is composed of a group of minicolumns, only one minicolumn is tuned to a particular feature that could occur in its receptive field and the input patterns are manually fed to different columns. Therefore, due to unused minicolumns, the memory capability of the model and its scaling up ability could be limited. Although the timing of spikes are employed to make the decision that which candidate is most likely to be the input pattern, less attention has been paid to the issue that how temporal information is exploited in the model.

We propose an hierarchically organized memory (HOM) model aiming to investigate the formation of neural cliques and the organization principle followed by learning with precisely timed spikes in a hierarchically structured spiking neural network. Sensory information travels upwards along the hierarchical network during the bottom-up information processing (Fig. 7.1). With a spike-timing based learning algorithm during the storing phase, the spiking neural network is able to map sensory information into neural clique activities. Meanwhile, auto-associative memory can be generated via fast NMDA dominated STDP process in Layer I, and sequence learning can be performed by slow NMDA dominated STDP in Layer II. The main contribution of this work is to propose a cognitive memory model focusing on memory organization using temporal population codes. We believe that the HOM model is a good attempt to explore the underlying mechanisms of formation and organization of memories in the brain.

The structure of this chapter is as follows: in the next section, we introduce the general structure of the HOM model, neural representation of sensory information, learning algorithm and etc. In Sect. 7.3, the performance of the proposed model is demonstrated by numeric simulation results. In Sect. 7.4, important issues of the HOM model and relations to other methods are discussed. At last, a summary of results is presented in Sect. 7.5.

7.2 The Hierarchical Organized Memory Model

The basic HOM model is composed of three layers: input layer, Layer I and Layer II. As shown in Fig. 7.1, neurons in the lower layer are fully connected to its next higher layer, whereas lateral connections exist in Layer I and Layer II. The interneurons aside provide feedback inhibition to prevent continuous firing and temporally separate firing events representing different memory items into gamma cycles.

7.2.1 Neuron Models and Neural Oscillations

The spike response model (SRM) [24], which provides a simple description of the spiking neuron, has been widely used in various studies. Pyramidal cells are employed in Layer I simulating short-term memory (STM). By utilizing the slow build-up ramp of after-depolarizing potential (ADP) of pyramidal cells [25], the status of neurons can be maintained. We plug ADP kernel into the generic spike response model, so that the state of pyramidal neuron i at time t is described as

$$V_i(t) = \eta_{ADP}(t - t_i) + \sum_j w_{ij}\varepsilon_{ij}(t - t_j) + V_{rest} \qquad (7.1)$$

where t_i and t_j denote firing times of the presynaptic neuron i and the post-synaptic neuron j, respectively. $\eta_{ADP}(t - t_i)$ is the ADP of neuron i, and w_{ij} is the synaptic efficacy from neuron j to neuron i. A spike is generated when the membrane potential reaches its threshold $V_{thr} = 1$. The membrane potential is set to $V_{rest} = 0$ after each firing, while the ADP is reset and build up again.

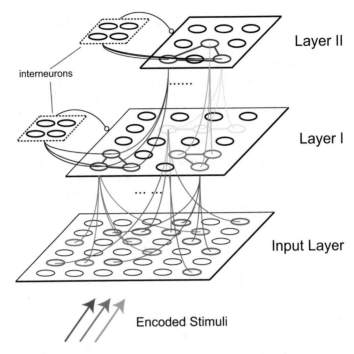

Fig. 7.1 The architecture of the three-layer HOM model. Neurons forming neural cliques and enhanced lateral connections are illustrated in different colors. Two groups of interneurons feed inhibition *back* to neurons in the same layer

The response of pyramid neuron after firing is modeled as an α function as follows:

$$\eta_{ADP}(t - t_i) = A_{ADP}\frac{t - t_i}{\tau_{ADP}} \exp(1 - \frac{t - t_i}{\tau_{ADP}}) \tag{7.2}$$

where A_{ADP} is the amplitude of ADP, and τ_{ADP} is the time constant affecting the duration of excitatory ramp.

Theta and gamma oscillations are two important types of brain wave in synchronizing the neural activity [9, 26]. They are believed to be critical for temporal coding/decoding of active neuronal ensembles, learning and memory formation [27–29]. An external theta oscillatory source, which injects current to neurons in Layer I, is modeled as a cosine wave

$$i_\theta = A_\theta \cos(2\pi f_\theta + \phi_0) \tag{7.3}$$

where A_θ is the amplitude of sub-threshold membrane potential oscillation, f_θ is the frequency of theta oscillation, and ϕ_0 is the initial phase. It has been found that the memory capacity depends on the theta/gamma cycle length ratio, suggesting that short-term memory is reserved within individual gamma cycles [30]. As the recent direct evidence from human EEG scalp recordings supports the STM model [31], each memory item is represented by firings in different gamma cycles in response to specific stimulus. Meanwhile, inhibition from interneurons suppresses individuals of other neural cliques.

7.2.2 Temporal Population Coding

With the interests in the functional role of temporal information carried by neural activities, temporal codes have received increasing attention [6, 32]. The information about stimulation is thought to be encoded by the time of spikes generated by a specific population of neurons, and each input pattern is coded by a particular group of neurons. Grayscale images are converted into single-spike spatiotemporal patterns with temporal population codes mimicking the visual sensory encoding. A original image is fed into Garbor filters and the output of Garbor filters are converted into neural firings according to the following equation.

$$t_i = f(s_i) = t_{max} - ln(\alpha \cdot s_i + 1) \tag{7.4}$$

where t_i is the firing time of neuron i, t_{max} is the width of encoding window, α is a scaling factor, and s_i is the intensity of output of Garbor filter (see Fig. 7.2). As a result, each spike codes orientation components of the image and the latency denotes the weight of corresponding component.

7.2.3 The Tempotron Learning and STDP

The learning rules employed in the model are expected to be compatible with temporal codes. Among the existing spike-timing based learning approaches, the tempotron rule has been shown to be a biologically plausible supervised synaptic learning scheme [17].

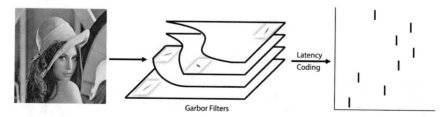

Garbor Filters

Fig. 7.2 Encoding scheme. A *grayscale* image is convolved with Garbor filters to extract orientation related features and then converted into a spike pattern by latency coding method

The tempotron learning has been employed in adjusting neural connectivities between layers. When presented a pattern, each neuron needs to make a decision on that whether the stimulus contains certain features that have been learned before. The connections from neurons that contribute to the integrated postsynaptic membrane potential during the presentation will be enhanced according to the tempotron learning rule (Eq. 7.5).

$$\Delta w_i = \lambda d \sum_{s_i < 0} exp(s_i) \tag{7.5}$$

where w_i is the synaptic weight from afferent i to the postsynaptic neuron, λ is the learning rate, d is the desired output label (either 0 or 1), and $s_i = t_i - t_{max}$ is the delay between presynaptic firing (S_i) and the time when postsynaptic membrane potential $V(t)$ reaches its maximal value V_{max}. The tempotron learning rule is illustrated in Fig. 7.3.

Because NMDA receptor (NMDAR) is the predominant molecular device for controlling synaptic plasticity, synaptic modifications (LTP and LTD) varies with different postsynaptic NMDARs [33, 34]. Although the biophysical and biochemical mechanisms that underpin STDP still need further investigation, these existing results suggest that STDP is a NMDAR-dependent mechanism. In the proposed model, fast and slow NMDA channels (fast: τ 25 msec, slow: τ 150 msec) are adopted to dominate the synaptic transmission and plasticity in Layer I and Layer II, respectively. The repetitive firings contribute to the enhancement of connections between activated neurons via STDP learning. As the time course of the activation of NMDARs crucially affects long term modification, STDP learning performs differently with NMDA channel in different states.

One should also note that feedback inhibition from interneurons leads to the restult that spike volleys are temporally separated into individual gamma cycles. Therefore, fast NMDA channel contributes to the formation of intra-clique connections (auto-associative memory) in Layer I, and slow NMDA channel spanning over several gamma cycles enhance inter-clique connections (episodic memory) in Layer II.

7.3 Simulation Results

In this section, we demonstrate that the proposed HOM model is capable of learning input patterns, storing them into individual gamma cycles with population firings, and performing sequence learning. Several experiments are conducted to illustrate and analyze these processes.

As shown in Fig. 7.1, the spiking neural network used to implement the HOM model is composed of three layers. The synaptic weights are initialized according to the population size of each layer.

Each input pattern is represented by tens of spikes using temporal population codes as shown in Fig. 7.2, and they are introduced to the network during troughs

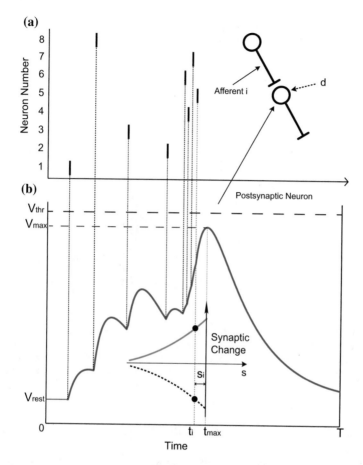

Fig. 7.3 Illustration of the tempotron learning rule. **a** Typical input pattern using temporal population codes. **b** Membrane potential of the postsynaptic neuron. The maximum value of the membrane potential is reached at t_{max}. (inset) The synaptic weight w_i changes accordingly to the time difference between s and the desired signal d. If $d = 1$, $\Delta w_i \geq 0$ (*solid line*), or if $d = -1$, $\Delta w_i < 0$ (*dashed line*)

of the theta oscillation. Fast NMDA channels maintain activated state around 10 ms after the binding of glutamate to postsynaptic cells, while slow NMDA channels with a slow deactivation time constant dominate the STDP process in Layer II (Fig. 7.4). The inter-layer synaptic weights are updated according to the tempotron learning rule during the representation of input patterns (gray strips in Fig. 7.5), while intra-layer synaptic plasticity are modified by STDP.

Driven by input synaptic currents from afferents, increasing number of pyramidal neurons in Layer I start to fire and form different neural cliques iteration by iteration. When enough stimulation are generated by neural activities in Layer I, similar phenomena emergence in Layer II. After dozens of iterations, neural cliques response

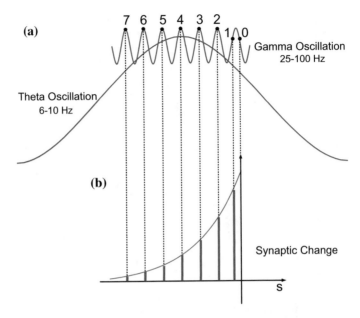

Fig. 7.4 LTP induced by STDP learning for different memory items. **a** Firings within each gamma cycle represent memory items 0–7. **b** Synaptic changes depend on the relative time between firings. Strong connections within the same neural cliques can be formed via the fast NMDA channel (1 → 0), while weaker LTPs are induced between neural cliques in different gamma cycles via the slow NMDA channel (7 → 0)

to specific patterns and repeat to fire periodically during the subsequent theta cycles (Fig. 7.5).

As shown in Fig. 7.5, we can identify groups of neurons firing as neural cliques in Layer I and II, respectively (Fig. 7.5b, c). Within each theta cycle, neural cliques respond selectively and repetitively to the stimulation as the same order that input patterns are introduced to the network. As can be seen from Fig. 7.5b, individual letters ('L', 'O', 'V' and 'E') are separately encoded by the volley activities of corresponding neural clique in Layer I. While neural activities generated by all the four neural cliques in Layer II are coding for the word 'LOVE'. The memory coding principle is explained in detail in section III A and B, respectively.

Figure 7.6 reveals the mechanism underlying repetitive firing of pyramidal neurons. After generation of the first spike by a particular neuron, its ADP starts to build up. When the slowly ramping up ADP meets near-peak theta current, the pyramidal neuron will fire again in the following theta cycle. Meanwhile, inhibitory feedback from interneurons prevents neurons coding for other patterns from firing right after the volley spikes. As a result, spike volleys are temporally separated into individual gamma cycles (Fig. 7.5a).

When neurons initially start to fire, the spike times are randomly distributed in gamma cycles. With repetitive firings, synaptic weights between individual members

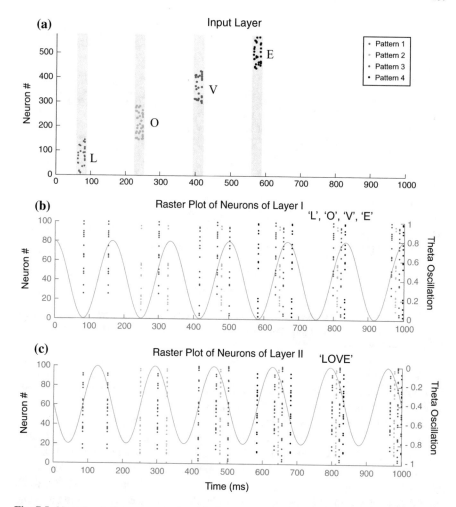

Fig. 7.5 Neural activity propagates through the system. **a** Encoded input patterns. Each pattern consists of tens of neurons firing within an encoding window (*gray strips*). **b** and **c** are the raster plots of the neural activities in Layer I and II, respectively. *Colored dots* denote spikes generated by neurons coding for different input patterns

of the same neural clique are strengthened with STDP learning, resulting in their synchronized firing as shown in Fig. 7.5b.

Since fast and slow NMDA channels dominate the STDP process in Layer I and II respectively, the resulting lateral connectivities are quite different. To verify it, we examine the synaptic weights, especially intra-clique connections in Layer I and inter-clique connections in Layer II as presented in Figs. 7.7 and 7.8.

When exposed to external stimuli, neurons in Layer I start to fire due to the enhancement of connections from input layer to Layer I as shown in Fig. 7.7. Once

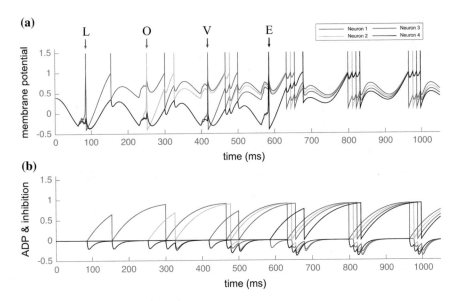

Fig. 7.6 Typical neural responses of pyramidal neurons in the same layer. **a** Membrane potentials of neurons coding for different patterns. **b** Slow built-up ADP of pyramidal neurons (positive) and inhibition from interneurons (negative)

neurons in Layer II receive enough stimulation from Layer I, they begin to generate spikes. At the same time, activated neurons within the same layer wire together to form neural cliques as shown in Figs. 7.7b (W_{22}) and 7.8b (W_{33}).

To further study the resulted neural cliques and their connectivities, we take a close look at synaptic connections within Layer I and II. Generally, lateral connections fall into intra-clique, inter-clique and weak connections. The connectivity developed after learning is illustrated in Fig. 7.9. As non-activated neurons wires weakly to all the other neurons, only intra-clique and inter-clique connections are drawn in the following figure.

Since fast NMDA channels stay activated for several milliseconds, only firings of neurons forming the same clique fall within this narrow time window. Consequently, intra-clique connections are enhanced via STDP process and salient as shown in Fig. 7.7b. The highlighted weights matrices show that each neural assembly forms a recurrent subnetwork with auto-associative memory coded in the enhanced lateral connections.

Although lateral connections in Layer II were strengthened as those in Layer I, the resulting connectivity is different from that in Layer I. As slow NMDARs have a longer activation period spanning over several gamma cycles, spikes in different cycles would induce enhancement of inter-clique connections. The salient weight elements along the diagonal are similar to those in Layer I, in which auto-associative memory is stored. While elements in the colored boxes denote connections between neural cliques, in which episodic memories are encoded.

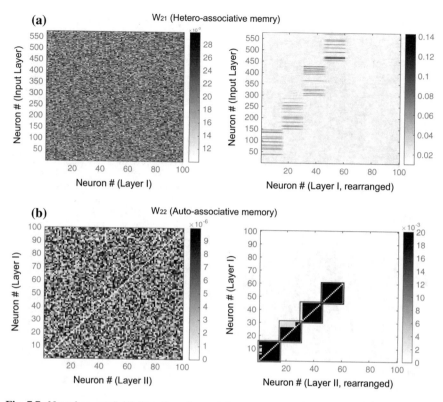

Fig. 7.7 Neural connectivity from input layer to Layer I (**a**) and within Layer I (**b**) before (*left*) and after learning (*right*). The activated neurons are picked out and rearranged for clear illustration in the *right column*. Intra-neural clique connections are highlighted by *colored boxes*

In sum, neurons forming the same neural clique tend to fire in synchrony, while neural cliques coding for successive patterns are temporally compressed. The former is caused by fast NMDA mediated STDP in Layer I, while the latter is a result of slow NMDA mediated STDP in Layer II. As demonstrated in Fig. 7.5, neurons coding for different memories fires in different gamma cycles in Layer I represent the detection of individual patterns, while neural responses within each theta cycle in Layer II representing the recognition of sequence of patterns. To understand from another angle, the neural cliques identified in Layer II can be considered as a assembly coding for the particular combination of patterns with their presentation sequence encoded. Therefore, hetero-association between layers abstract information during the bottom-up process, while inter-clique episodic memory binds information about several temporally separated patterns (individual letters) into a compressed pattern (a complete word).

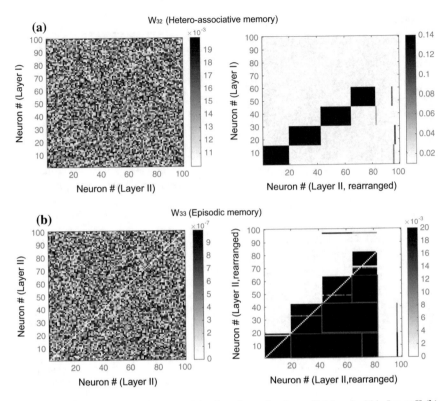

Fig. 7.8 Evolution of the neural connectivity from Layer I to Layer II (**a**) and within Layer II (**b**). Inter-neural clique connections in Layer II are highlighted by *colored boxes*

7.3.1   Auto-Associative Memory

Our brain has a remarkable ability of association, despite constant changes in real-world circumstance. During perception along sensory pathways, information about external stimulation is abstracted and encoded into reliable neural activities. After training, both hetero-associative memory and auto-associative memory are stored in the connections between neurons. Input patterns are hetero-associated with neural responses in Layer I via synaptic weights between input layer and Layer I, while auto-associative memory is represented by intra-clique connections.

As neural activities can be observed as an explicit expression of stored memory, pattern completion may refer to the ability that a subset of neurons from a particular neural clique are able to arouse the rest of that clique. The trained network is expected to be competent for recalling stored neural activities upon presentation of input patterns and retaining invariant responses in presence of noises and even corruption of information. As information is distributed over neurons with population coding scheme, the information loss caused by shifting or removing spikes

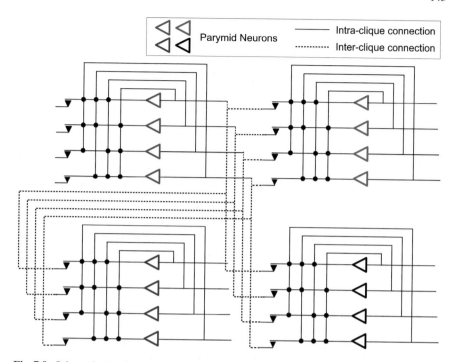

Fig. 7.9 Schematic diagram of developed lateral connectivity. Lateral connections within the same layer are divided into intra-clique and inter-clique connections. Only inter-clique connections from neural clique 1 to the rest are drawn for clear illustration

can be complemented with the aid of other contributing neurons. In order to investigate this capability of reproducing neural activities, time jitter and missing of spikes are considered in the following experiments. Hence, a correlation-based measure of spike timing [35] is used to calculate the distance between an output pattern and its corresponding target pattern.

$$C = \frac{\overrightarrow{s_1} \cdot \overrightarrow{s_2}}{|\overrightarrow{s_1}||\overrightarrow{s_2}|} \tag{7.6}$$

where C is the correlation denoting closeness between two temporal coded patterns (s_1 and s_2). They are convolved with a low pass Gaussian filter of a width $\sigma = 2$ ms.

By shifting firing times of input spikes, variability of input patterns was simulated as shown in Fig. 7.10a. The shifting intervals were randomly drawn from a gaussian distribution with mean 0 and variance $[0, 5]$ ms. The correlation between reproduced neural responses and the desired patterns are presented in Fig. 7.10. Every simulation has been repeated for 30 times to generate the averaged performance. Figure 7.10b shows that the network reproduces reliable neural patterns in the presence of shifting of input spikes up to 3 ms in both Layer I and II. However, the performance dramatically drops to around 0.3 as the shifting interval increases to 5 ms. Neural responses

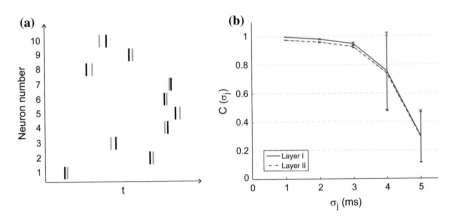

Fig. 7.10 a Illustration of shifted spatiotemporal patterns. Firing times of original input spikes (*black* bars) are randomly shifted with random jitters (*gray* bars). **b** Reliability of retrieved neural responses under different noise levels

in Layer I are slightly more robust than those in Layer II due to error accumulation during the upwards information transmission.

An additional experiment is conducted to further investigate the link between intra-clique connections and auto-associative memory. Since neurons in the same clique may provide supplementary stimulation to sustain an united activity, corruption of input patterns may not be a catastrophic error. All settings are the same as the previous experiment, whereas one out of ten spikes is removed from each training pattern. The experiment has been repeated for 20 times and the mean value of the correlation between actual output and desired pattern is calculated for each trial.

As illustrated in Fig. 7.11a, intra-clique connections are enhanced during learning, while non-selective neurons are weakly connected to all the other neurons. To verify that connections within neural cliques are responsible for the completion of patterns in Layer I, lateral connection are removed. As shown in Fig. 7.11b, the experimental results are consistent with our analysis.

Knowledge stored in the synaptic weights from the input layer to Layer I provides the capability of hetero-association (Fig. 7.7a) by recognizing input patterns via their specific features. At the same time, intra-clique connections (Fig. 7.7b) contribute to the pattern completion which is one of the most important features of auto-associative memory, in Layer I. Since corrupted patterns provide insufficient stimulation to neurons in the next layer, some of the trained neurons that should have been activated may not be triggered. Fortunately, lateral inputs from excited neurons can provide supplementary information to recall desired neural responses. Therefore, associative memory, which can be understood as the ability to retrieve invariant responses with partial information, relies on both the distributed storage of knowledge in synaptic connectivity between layers and within Layer I.

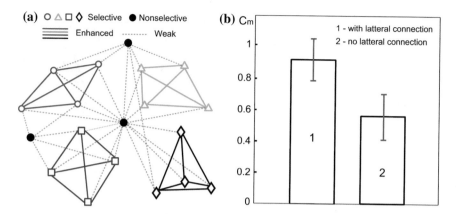

Fig. 7.11 a Illustration of neural cliques in Layer I coding for different input patterns (letters) after learning. Intra-clique connections of selective neurons are enhanced to form recurrent networks and non-selective neuron are weakly connected to other neurons. **b** Test results of the associative memory based on the correlation between retrieved and corresponding desired patterns in response to corrupted input patterns

7.3.2 Episodic Memory

In the previous experiment, fast NMDA channel contributes to the formation of neural clique together with the auto-associative memory. While slow NMDA channel dominates the STDP process in Layer II, which leads to different postsynaptic neural responses and different connectivity. The slow decaying time constant of slow NMDA channel leads to accumulation of EPSPs from different neural cliques. Meanwhile, slow NMDA receptors sustain activation state over several gamma cycles, which enables STDP learning to link sequence of memory items by building up inter-clique connections. When lateral connections are sufficiently developed, the accumulated EPSPs of occurred memory items would be able to trigger subsequent items without the presentation of expected upcoming input stimulation during memory recall.

As demonstrated in Fig. 7.12, stimulation caused by neural cliques coding for the first three memory items in the sequence is strong enough to trigger the particular neural clique coding for the missing item. The inter-clique connections may lead to the result that consecutive memory items are temporally compressed as a group of neuron coding for the combination of several patterns in the sequence. This characteristic is crucial for a spike-timing based hierarchical model, which contributes to pattern/information binding process.

Figure 7.13 shows how EPSPs of consecutive memories lead to the activation of neural cliques coding for the next upcoming pattern, which is not presented to the network. Since only the first three letters were inserted during recall of episodic memory, neural cliques coding for them in Layer I and Layer II fired in response to the stimulation as shown in Fig. 7.13a, c. However, due to the slow-NMDA mediated synaptic transmission and enhanced inter-clique connections in Layer II, neural

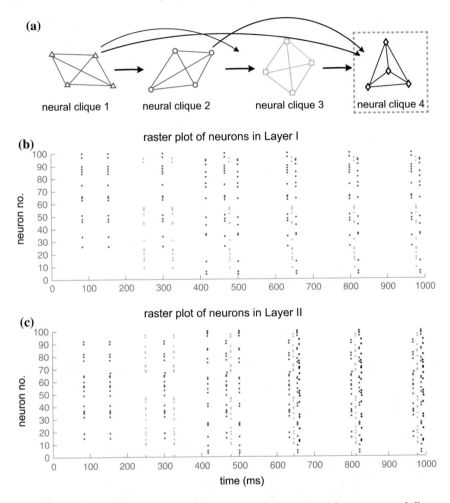

Fig. 7.12 **a** Illustration of generated connectivity in Layer II. Synaptic weights among neural cliques are strengthened via slow NMDA dominated STDP learning. **b** Raster plot of neural activities in Layer I during recall. Neurons coding for letters 'L', 'O' and 'V' detect the presentation of them and trigger corresponding firings in Layer II. **c** Episodic memory stored in Layer II enables the recall of missing item ('E') by triggering firing of neurons coding for it

clique coding for the next "missing" pattern were triggered by the accumulation of EPSPs induced by the neural cliques coding for preceding patterns (Fig. 7.13b). As a result, we can retrieve the full sequence by observing the neural activities in Layer II as shown in Fig. 7.13c. In addition, neural activities can be converted to binary codes according to their behaviors within a certain coding time window (gray strips in Fig. 7.13c). By reading out these binary codes, we can identify the presence of individual features/patterns in Layer I and combination of features/patterns (sequence) in Layer II.

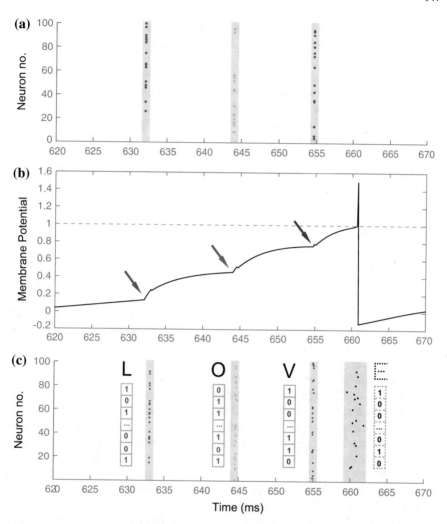

Fig. 7.13 Recall of neural clique activities induced by accumulated EPSPs. **a** Response of neural cliques coding for the presented patterns in Layer I. **b** Membrane potential of an activated neuron coding for the expected upcoming pattern ('E') in Layer II. **c** Raster plot of neural activities in Layer II and their corresponding binary codes

7.4 Discussion

7.4.1 Information Flow and Emergence of Neural Cliques

Basically, the information flow in the model is unidirectional from bottom to top. Information between layer travels upward along the network with a filtering process, while recurrent subnetworks (neural cliques) exist in both Layer I and Layer II. Input stimulation trigger repetitive neural activities in Layer I, these activities further drive neural responses in Layer II. In addition, interneurons aside each layer feed inhibition back to neurons that trigger them.

During learning, neurons compete with each other to respond selectively to specific stimulus. Synaptic weights between layers stop changing when the population sizes of evoked neurons reach predefined configuration. Since neural responses in higher layer rely on the input stimulation from lower layer, the emergence of neural cliques in higher layer lags those in lower layer. Although bounded synapses have limited memory storage capacity [36], they are used in our model to ensure a certain number of presynaptic neurons contributing the generation of postsynaptic spikes.

7.4.2   Storage, Recall and Organization of Memory

After memory representation is defined as neural clique response, storage and recall are the rest two key issues for a memory model. During learning input patterns, hetero-association is achieved by enhanced synaptic weights between layers. Once pyramidal neurons are triggered to fire, cooperation of ADP and theta oscillation results in the repetitive firing of neurons coding for memory items every theta cycle. While STDP learning with fast/slow NMDA channels contributes to the enhancement of intra- and inter-clique connections, respectively.

Theta oscillation has been recorded in hippocampus involved of memory function. A recent model [37] suggests that memories might be encoded and recalled during different portions of the theta cycle. Similar scheme is employed in our model, hetero-associative memory storage occurs during troughs of theta oscillation, while stored memories are retrieved during portions near the maximums of theta oscillation. Since activation of neural activities at troughs requires strong excitation, the resulted synaptic efficacies are stronger than required at the maximums. Redundant excitation and distributed information over neurons improves the robustness of recalling hetero-associative memories. Environmental noises and even information loss will not lead to a catastrophic retrieval failure as demonstrated in the simulation results. In addition, inhibition of negative half of theta oscillation suppresses excitation induced by enhanced lateral connections when storage. On the other hand, if insufficient stimulation is injected into the network during retrieval, excitation of positive half may provide supplementary stimulation.

As demonstrated in simulation results, multiple patterns are encoded and stored into associative and episodic memory following a hierarchical organization principle. Hereto-associative memory is encoded by the connectivity between layers, which plays a pivotal role in recognizing specific patterns. Along with the development of neural cliques, lateral connections are enhanced by STDP process in Layer I and II. Intra-clique connections represent auto-associative memory, while episodic memory about the sequence of input patterns are encoded in the form of inter-clique connections.

7.4.3 Temporal Compression and Information Binding

The discovery of place cells suggests that spatial information can be encoded by the cellular activities of hippocampus in the brain. Moreover, dual oscillations have been observed in the brain involved in memory function. In the HOM model, memory items are coded by neuron assemblies firing within different gamma cycles, while past and present events are chunked by the theta oscillation. When arriving a particular place that has been visited before, neural clique coding for it will be activated. Later when coming up with another landmark, another group of neurons will be activated. The temporal compression of neural firing volleys contributes to the generation of inter-clique connections and ability to predict upcoming patterns. Due to the repetitive firing of neurons coding for the previous and current landmarks, the superimposed impact of these neural cliques will trigger the third clique, predicting the upcoming landmark in the learned sequence.

Since neural responses in Layer II are linked with inter-clique connections, information about different stimuli is binded. As shown in Fig. 7.5c, neural activities of different cliques in Layer II are interconnected. Therefore, temporally compressed neural patterns can be understood as a new neural clique in Layer II, and fed into other basic three-layer networks as input. By duplicating this basic network into a larger network with hierarchical structure, more powerful ability to organize neural activities representing features with different specificity along the hierarchy can be achieved. Each basic networks binds several features into a combined feature and transmits it to higher level network as its input stimulus. Therefore, the neural cliques representing the most general features are at the bottom. Moving up along the hierarchy, neural activities represents more specific and complex patterns.

7.4.4 Related Works

Jensen and colleagues have demonstrated that simultaneous firings of a group of neurons can be stored in a fixed recurrent network modeling hippocampual CA3 area [38]. The idea that dual oscillation interacts with pyramidal neurons has been implemented in the model. Although firing times of spike are considered in the model, the external inputs exciting specific pyramidal neuron are presumed fire in synchrony, which ignores sensory encoding as well as the hetero-association process. In addition, recurrent subnetworks are predefined in the model and input patterns are presented to specific recurrent networks. These assumptions restrict the generalization and adaptability of the proposed model.

The *Cortext* model focuses on hetero-associative memory for word recognition, while auto-associative memory is implemented in Jensen et al.'s CA3 model. Similar to them, most existing memory models consider specific memory function and rarely consider the memory organization issue. In contrast, both associative memory (including hetero- and auto-associative) and episodic memory are involved in HOM model. The functional roles are clearly assigned and they have been hierarchically organized. Hereto-associative memory encoded by inter-layer connections to trigger neurons in upper layers, auto-associative memory is encoded by intra-clique connections to complete corrupted spatial temporal patterns, and episodic memory is stored in intra-clique connections to evoke upcoming memories in a sequence during recall.

In contrast to fixed structures, the HOM model can generate neural cliques during the learning process. The plasticity of network structure not only provides generalization and scale-up capability, but fully exploit available coding units of the network. Therefore, the proposed network architecture is a structure that evolves to a complex connectivity after training and adapt accordingly to specific input stimuli.

From the perspective of biological plausibility, all information processing phases in HOM purely rely on the firing time of spikes. Temporal population coding scheme together with spike-timing based learning are employed to investigate how time information might be exploited in memory system. Spatiotemporal patterns are fed to model sequentially as inputs and neural cliques activities can be observed as internal representation of memory items. In addition, the HOM model employ spike-timing dependent learning schemes to adapt both inter-layer connections and intra-layer connections.

One should note that the proposed HOM model share some similar ideas with several existing studies. The HOM model simulates neural clique activities reported in [14]. We agree with the separation of encoding and retrieval theory suggested by [37, 39]. The mechanism sustains STM was used in Jensen et al.'s model, while tempotron learning and STDP learning have be employed in contributing long-term memory formation in other models. However, some important issues need to be further studied. Since bounded synapses are used in the model, overlapping of neural cliques and soft limits of weight may contribute to improve the memory capacity of the model. In addition, more investigation is needed to study how to organize similar features together with a more complex model by duplicating and connecting several basic networks together.

7.5   Conclusion

In this chapter, a hierarchically organized memory model is described, demonstrated and analyzed. The HOM model advances our understanding of the relationship between spike-timing based learning and formation of memory. Dual oscillation theory has been applied to achieve short-term memory. Fast NMDA channel, which activates within the following gamma cycle, contributes to the formation of auto-associative memory in Layer I. While slow NMDA channel spanning over several gamma cycles enables the synaptic facilitation among different neural cliques, resulting in the development of episodic memory. In addition, individual patterns can be recognized by observing neural responses of neural cliques in Layer I, while sequence of them can be detected by observing the collective activities in Layer II.

We believe that this is a good attempt to investigate memory generation and organization with temporal based coding and learning schemes in spiking neural network. Different functional memory types are involved in the model and combination of memories occurs in the top layer as demonstrated. With a more complicated network, the capability of generating neural cliques representing memories with different specificity in hierarchy can be further explored. Moreover, real-world stimuli such as visual and auditory signal can be employed as the sensory information to investigate the application potential of the HOM model. We hope that our study can advance our understanding of the basic memory organizing principles and contributes to the development of artificial cognitive memory.

References

1. Meister, M., Berry II, M.J.: The neural code of the retina. Neuron **22**(3), 435–450 (1999)
2. Heil, P.: Auditory cortical onset responses revisited. i. first-spike timing. J. Neurophysiol. **77**(5), 2616–2641 (1997)
3. Perez-Orive, J., Mazor, O., Turner, G.C., Cassenaer, S., Wilson, R.I., Laurent, G.: Oscillations and sparsening of odor representations in the mushroom body. Science **297**(5580), 359–365 (2002)
4. Mehta, M.R., Lee, A.K., Wilson, M.A.: Role of experience and oscillations in transforming a rate code into a temporal code. Nature **417**, 741–746 (2002)
5. VanRullen, R., Guyonneau, R., Thorpe, S.J.: Spike times make sense. Trends Neurosci. **28**(1), 1–4 (2005)
6. Kayser, C., Montemurro, M.A., Logothetis, N.K., Panzeri, S.: Spike-phase coding boosts and stabilizes information carried by spatial and temporal spike patterns. Neuron **61**(4), 597–608 (2009)
7. Samonds, J.M., Zhou, Z., Bernard, M.R., Bonds, A.B.: Synchronous activity in cat visual cortex encodes collinear and cocircular contours. J. Neurophysiol. **95**(4), 2602–2616 (2006)
8. Georgopoulos, A., Schwartz, A., Kettner, R.: Neuronal population coding of movement direction. Science **233**(4771), 1416–1419 (1986)
9. O'Keefe, J., Dostrovsky, J.: The hippocampus as a spatial map: preliminary evidence from unit activity in the freely-moving rat. Brain Res. **34**, 171–175 (1971)
10. Leutgeb, S., Leutgeb, J., Moser, M.B., Moser, E.: Place cells, spatial maps and the population code for memory. Curr. Opin. Neurobiol. **15**(6), 738–746 (2005)
11. Wyss, R., König, P., Verschure, P.F.M.J.: Invariant representations of visual patterns in a temporal population code. Proc. Natl. Acad. Sci. **100**(1), 324–329 (2003)
12. Lin, L., Osan, R., Shoham, S., Jin, W., Zuo, W., Tsien, J.Z.: Identification of network-level coding units for real-time representation of episodic experiences in the hippocampus. Proc. Natl. Acad. Sci. **102**(17), 6125–6130 (2005)
13. Kiani, R., Esteky, H., Mirpour, K., Tanaka, K.: Object category structure in response patterns of neuronal population in monkey inferior temporal cortex. J. Neurophysiol. **97**, 4296–4309 (2007)
14. Lin, L., Osan, R., Tsien, J.Z.: Organizing principles of real-time memory encoding: neural clique assemblies and universal neural codes. Trends Neurosci. **29**(1), 48–57 (2006)
15. Durstewitz, D., Seamans, J.K., Sejnowski, T.J.: Neurocomputational models of working memory. Nature Neurosci. **3**, 1184–1191 (2000)
16. Tang, H., Ramanathan, K., Ning, N.: Guest editorial: special issue on brain inspired models of cognitive memory. Neurocomputing **138**, 1–2 (2014)

17. Gütig, R., Sompolinsky, H.: The tempotron: a neuron that learns spike timing-based decisions. Nature Neurosci. **9**(3), 420–428 (2006)
18. Bohte, S.M., Bohte, E.M., Poutr, H.L., Kok, J.N.: Unsupervised clustering with spiking neurons by sparse temporal coding and multi-layer RBF networks. IEEE Trans. Neural Netw. **13**, 426–435 (2002)
19. Ponulak, F., Kasinski, A.: Supervised learning in spiking neural networks with resume: sequence learning, classification, and spike shifting. Neural Comput. **22**(2), 467–510 (2010)
20. Florian, R.V.: The chronotron: a neuron that learns to fire temporally precise spike patterns. PLoS One **7**(8), e40233 (2012)
21. Hu, J., Tang, H., Tan, K.C., Li, H., Shi, L.: A spike-timing-based integrated model for pattern recognition. Neural comput. **25**(2), 450–472 (2013)
22. Yu, Q., Tang, H., Tan, K.C., Li, H.: Precise-spike-driven synaptic plasticity: learning hetero-association of spatiotemporal spike patterns. PLoS One **8**(11), e78318 (2013)
23. Schrader, S., Gewaltig, M.O., Körner, U., Körner, E.: Cortext: a columnar model of bottom-up and top-down processing in the neocortex. Neural Netw. **22**(8), 1055–1070 (2009)
24. Maass, W., Bishop, C.M.: Pulsed Neural Networks. MIT Press, Cambridge (2001)
25. Jensen, M.S., Azouz, R., Yaari, Y.: Spike after-depolarization and burst generation in adult rat hippocampal ca1 pyramidal cells. J. Physiol. **492**, 199–210 (1996)
26. Nicolelis, M., Baccala, L., Lin, R., Chapin, J.: Sensorimotor encoding by synchronous neural ensemble activity at multiple levels of the somatosensory system. Science **268**(5215), 1353–1358 (1995)
27. Buzsáki, G.: Theta oscillations in the hippocampus. Neuron **33**, 325–340 (2002)
28. Lega, B.C., Jacobs, J., Kahana, M.: Human hippocampal theta oscillations and the formation of episodic memories. Hippocampus **22**, 748–761 (2012)
29. Axmacher, N., Mormann, F., Fernández, G., Elger, C.E., Fell, J.: Memory formation by neuronal synchronization. Brain Res. Rev. **52**(1), 170–182 (2006)
30. Lisman, J.E., Idiart, M.A.: Storage of 7 +/- 2 short-term memories in oscillatory subcycles. Science **267**(5203), 1512–1515 (1995)
31. Kamiński, J., Brzezicka, A., Wróbel, A.: Short-term memory capacity (7+/-2) predicted by theta to gamma cycle length ratio. Neurobiol. Learn. Mem. **95**(1), 19–23 (2011)
32. Singer, W., Gray, C.M.: Visual feature integration and the temporal correlation hypothesis. Annu. Rev. Neurosci. **18**(1), 555–586 (1995)
33. Bear, M.F., Malenka, R.C.: Synaptic plasticity: LTP and LTD. Curr. Opin. Neurobiol. **4**(3), 389–399 (1994)
34. Malenka, R.C., Bear, M.F.: LTP and LTD: an embarrassment of riches. Neuron **44**(1), 5–21 (2004)
35. Schreiber, S., Fellous, J., Whitmer, D., Tiesinga, P., Sejnowski, T.: A new correlation-based measure of spike timing reliability. Neurocomputing **52–54**, 925–931 (2003)
36. Fusi, S., Abbott, L.F.: Limits on the memory storage capacity of bounded synapses. Nature Neurosci. **10**(4), 485–493 (2007)
37. Hasselmo, M.E., Bodelón, C., Wyble, B.P.: A proposed function for hippocampal theta rhythm: separate phases of encoding and retrieval enhance reversal of prior learning. Neural Comput. **14**(4), 793–817 (2002)
38. Jensen, O., Idiart, M., Lisman, J.E.: Physiologically realistic formation of autoassociative memory in networks with theta/gamma oscillations: role of fast nmda channels. Learn. Mem. **3**(2–3), 243–256 (1996)
39. Kunec, S., Hasselmo, M.E., Kopell, N.: Encoding and retrieval in the ca3 region of the hippocampus: a model of theta-phase separation. J. Neurophysiol. **94**(1), 70–82 (2005)

Chapter 8
Spiking Neuron Based Cognitive Memory Model

Abstract Jensen et al. (Learn. Mem. 3(2–3), 245–246, 1996 [1]) proposed an auto-associative memory model using an integrated short-term memory (STM) and long-term memory (LTM) spiking neural network. Their model requires that distinct pyramidal cells encoding different STM patterns are fired in different high-frequency gamma subcycles within each low-frequency theta oscillation. Auto-associative LTM is formed by modifying the recurrent synaptic efficacy between pyramidal cells. In order to store auto-associative LTM correctly, the recurrent synaptic efficacy must be bounded. The synaptic efficacy must be upper bounded to prevent re-firing of pyramidal cells in subsequent gamma subcycles. If cells encoding one memory item were to re-fire synchronously with other cells encoding another item in subsequent gamma subcycle, LTM stored via modifiable recurrent synapses would be corrupted. The synaptic efficacy must also be lower bounded so that memory pattern completion can be performed correctly. This chapter uses the original model by Jensen et al. as the basis to illustrate the following points. Firstly, the importance of coordinated long-term memory (LTM) synaptic modification. Secondly, the use of a generic mathematical formulation (spiking response model) that can theoretically extend the results to other spiking network utilizing threshold-fire spiking neuron model. Thirdly, the interaction of long-term and short-term memory networks that possibly explains the asymmetric distribution of spike density in theta cycle through the merger of STM patterns with interaction of LTM network.

8.1 Introduction

A key functional role of the hippocampus is the storage and recall of associative memories [2, 3]. Auto-association refers to the retrieval or completing of a memory from a partial or noisy sample of itself. Hetero-association refers to the recall of a memory from one category as a result of a cue from another category. The CA1 region of the hippocampus has been proposed to be hetero-associator [4]. Depending on the kinetics of NDMA channels, CA3 region of hippocampus can function as either hetero-associator [5] or auto-associator [1] for the storage of declarative memories.

© Springer International Publishing AG 2017
Q. Yu et al., *Neuromorphic Cognitive Systems*, Intelligent Systems
Reference Library 126, DOI 10.1007/978-3-319-55310-8_8

153

Dual oscillations have been recorded in hippocampus in which a low frequency theta oscillation is subdivided into about seven subcycles of high frequency gamma oscillation [6]. The theta rhythm in the hippocampus refers to the regular oscillations of the local field potential at frequencies of 4–12 Hz which has been observed in rodents [7]. In humans, the theta rhythm typically refers to oscillations in the frequencies of 4–7 Hz [8], while gamma rhythm typically refers to oscillations in the frequencies of 25–100 Hz [9]. It is thought that different information can be stored at different phases of a theta cycle [10]. This type of neural information representation is commonly known as phase encoding. It is also proposed that the theta rhythm could work in combination with another brain rhythm known as the gamma rhythm, of frequencies 40–100 Hz [6], to actively maintain auto-associative memories [1, 11]. Theta rhythm may also have a role to play in the formation of a cognitive map in the hippocampus [12, 13].

In this chapter, the auto-associative memory model proposed by [1] is presented using the Spiking Response Model (SRM) neurons [14, 15]. With SRM, the characteristic of a neuron is defined by kernel functions which describe the responses to presynaptic spike as well as refractory function that characterizes its response after it has fired. In this way, the membrane potential of a neuron contributed by these functions at different time instances can separately be identified and analyzed.

The auto-associative LTM is formed by modifying the recurrent synaptic efficacy between pyramidal cells. If the weights are not upper bounded, pyramidal cells that have fired would be re-triggered into subsequent gamma subcycle by the spikes feedback via the recurrent synaptic connections between these cells. This will conflict with cells encoding another memory item that fire in that gamma subcycle, and subsequently corrupt the LTM stored in the recurrent collaterals. If the synaptic weights of the recurrent collaterals are not lower bounded, pattern completion can not be carried out.

In addition, the analysis on the synaptic bounds of the original model reveals addition evidence for rejecting the view that STM and LTM are two separate entities. The interaction between LTM and STM network can cause evenly distribution spike density in theta cycle to become asymmetric distributed similar to the experimental result in [16]. Persistent burst of one STM pattern is merged with subsequent memory pattern to result in uneven distribution of the number of simultaneous spikes to represent the memory patterns maintained in STM network. The merger of patterns is accomplished in two steps. Firstly, sufficient inhibition terminates the continual burst of the first group of cells via the activation of another subset of cells. Secondly, the after-depolarization (ADP) intrinsic to cells causes the two groups of cells to subsequently fire in synchrony.

This chapter uses the original model by [1] as the basic model to illustrate the following points.

- The importance of coordinated long-term memory (LTM) synaptic modification [17].

- The use of a generic mathematical formulation (spiking response model [14, 15]) that can theoretically extend the results to other spiking network utilizing threshold-fire spiking neuron model.
- The interaction of long-term and short-term memory networks that possibly explains the asymmetric distribution of spike density in theta cycle [16] through the merger of STM patterns with interaction of LTM network.

8.2 SRM-Based CA3 Model

An overview of the SRM-based auto-associative memory model of Jensen et al. [1] is shown in Fig. 8.1 (see [18] for more details). The recurrent network in Fig. 8.1 models the hippocampal CA3 region. The CA3 system operates as an auto-association network and provides for the completion of stored memories during recall from a partial cue via entorhinal cortex (EC) [19]. All pyramidal cells in the CA3 also accept an oscillatory input that is used to model the theta rhythm. Feedback inhibition from interneurons is applied to all pyramidal cells.

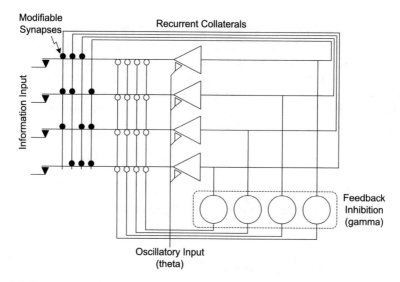

Fig. 8.1 Overview of CA3 model using SRM. The network consists of a short-term memory (STM) and long-term memory (LTM) network. STM repeats the memory items in every theta cycle. Interneurons provide feedback inhibition which generates the gamma subcycles within the positive portion of each theta cycle. LTM network encodes information in modifiable recurrent synapses. Adapted from [18]

8.2.1 Spike Response Model

Mathematically, the membrane potential of a neuron i under the SRM model is described by a state variable u_i. A spike is modelled as an instantaneous event that occurs when the membrane potential u_i exceeds a threshold V_{thres}. The time at which u_i crosses V_{thres} is said to be the firing time $t_i^{(f)}$. The set of all firing times of neuron i is denoted by

$$\mathscr{F}_i = \left\{ t_i^{(f)}; 1 \leq f \leq n \right\} = \left\{ t \,|\, u_i(t) = V_{\text{thres}} \wedge u_i'(t) > 0 \right\} , \qquad (8.1)$$

where n is the length of the simulation. After a spike has occurred at $t_i^{(f)}$, the state variable u_i will be reset by adding a negative contribution $\eta_i(t - t_i^{(f)})$ to u_i. The kernel $\eta_i(s)$, known as the refractory function, vanishes for $s \leq 0$ and decays to zero for $s \to \infty$. The refractory kernel defines a refractory period immediately following a spike during which the neuron will be incapable of firing another spike. The neuron may also receive input from presynaptic neurons $j \in \Gamma_i$ where

$$\Gamma_i = \{ j \,|\, j \text{ presynaptic to } i \} . \qquad (8.2)$$

A presynaptic spike increases or decreases the state variable u_i of neuron i by an amount $w_{ij}\varepsilon_{ij}(t - t_j^{(f)})$. The weight w_{ij} is known as the synaptic weight and it characterises the strength of the connection from neuron j to neuron i. The kernel $\varepsilon_{ij}(s)$ models the response of neuron i to presynaptic spikes from neurons $j \in \Gamma_i$ and vanishes for $s \leq 0$. In addition to spike input from other neurons, a neuron may receive external input h^{ext}, for example from non-spiking sensory neuron. Under the SRM model [14], the state $u_i(t)$ of neuron i at time t is hence modelled by (8.3).

$$u_i(t) = \eta_i \left(t - \hat{t}_i \right) + \sum_{j \in \Gamma_i} w_{ij}\varepsilon_{ij} \left(t - \hat{t}_j \right) + h^{\text{ext}}(t) , \qquad (8.3)$$

where $\hat{t}_i = t_i^{(n)} < t$ denotes the most recent firing of neuron i.

8.2.2 SRM-Based Pyramidal Cell

Using SRM, the dynamic of the pyramidal cells in Fig. 8.1 is modelled as follows. A pyramidal cell is fired when its membrane potential u_i^{PC} exceeds threshold $V_{\text{thres}}^{\text{PC}} = 10\,\text{mV}$ (see [20] for approximate value). In hippocampal pyramidal cells, action potentials are followed by after-hyperpolarizations (AHPs) [21, 22]. In addition, pyramidal cells exhibit an after-depolarization (ADP) after spike during cholinergic [21] or serotonergic modulation [23] or as a result of metabotropic glutamate receptors involved in the conversion of AHP to ADP [22]. Within the framework of the SRM neuron model, the refractory kernel $\eta_i(t - t_i^{(f)})$ of a pyramidal cell i is hence modelled to incorporate AHP and ADP (see (8.4)).

$$\eta_i^{PC}(s) = V_{refr}^{PC}(s) = \tag{8.4}$$

$$A_{AHP} \exp\left(-\frac{s}{\tau_{AHP}}\right) + A_{ADP}\frac{s}{\tau_{ADP}} \exp\left(1 - \frac{s}{\tau_{ADP}}\right),$$

where $A_{AHP} = -3.96\,\text{mV}$, $\tau_{AHP} = 5\,\text{ms}$ (see [21] for approximate value), $A_{ADP} = 9.9\,\text{mV}$, and $\tau_{ADP} = 200\,\text{ms}$ (see [22] for approximate value). The AHP prevents the pyramidal cells from fast repetitive firing whereas ADP provides a ramp of excitation that builds up after a spike within a time of 200 ms and then falls. Figure 8.2a illustrates the refractory kernel for the pyramidal cells.

Theta oscillations have been recorded in hippocampus [6]. Temporal correlations between active cells in medial septum and the hippocampal system indicate that the medial septum provides a constant cholinergic modulation that facilitates oscillations and induces a phasic drive [24]. Successful memory formation is correlated with tight coordination of spike timing with the local theta oscillation [25]. This external theta oscillatory signal is modelled by (8.5).

$$h^{ext}(t) = V_{theta}(t) = A_{theta} \sin(2\pi f t + \phi), \tag{8.5}$$

where $A_{theta} = 4.95\,\text{mV}$, $f = 6\,\text{Hz}$, and $\phi = -\pi/2$.

The spikes from EC or DG input serve to excite the CA3 pyramidal cells. The spikes from EC and DG layer are phase modulated by the local field potential with theta rhythm (LFP theta). The theta rhythm of EC has about $\pi/4$ phase shift from that of the LFP theta of the CA3 [26–29]. In this chapter, the LFP theta oscillation to the EC and DG layer is not modelled. For simplicity, the spikes from the input neurons are introduced into pyramidal cells at the troughs of the LFP theta cycles

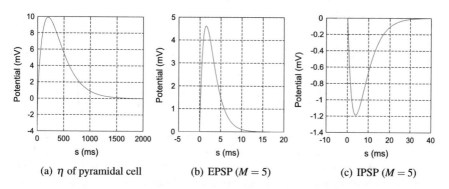

(a) η of pyramidal cell   (b) EPSP ($M = 5$)   (c) IPSP ($M = 5$)

Fig. 8.2 Kernel functions. **a** The refractory function of each pyramidal cell in which AHP proceeds before ADP after each spike of pyramidal cell. **b** The response of each pyramidal cell to each spike from other presynaptic pyramidal cells. **c** The response of pyramidal cell to each spike from the inhibitory interneurons

of CA3. In addition, the response $\varepsilon_{ij}^{PC\leftarrow I}$ of pyramidal cells i to presynaptic spikes from EC/DG input neuron j is modelled by (8.6). The response would ensure that membrane potential u_i^{PC} of pyramidal cell i reaches V_{thres}^{PC} at the troughs of the theta cycles. The synaptic transmission $w_{ij}^{PC\leftarrow I}$ from the input neuron j to pyramidal CA3 cell i is assumed with unit weight.

$$\varepsilon_{ij}^{PC\leftarrow I}(t - t_j^{(f)}) = V_{in}(t - t_j^{(f)}) = \tag{8.6}$$

$$\begin{cases} 14.982\,\text{mV} & \text{if } (t - t_j^{(f)}) = 0 \text{ ,} \\ 0\,\text{mV} & \text{if } (t - t_j^{(f)}) \neq 0 \text{ .} \end{cases}$$

For synaptic transmission between pyramidal cells in the recurrent collaterals, the response kernel $\varepsilon_{ij}^{PC\leftarrow PC}$ denotes an excitatory postsynaptic potential (EPSP) V_{EPSP}. The synaptic transmission is mediated by synaptically released glutamate binding to AMPA (α-amino-3-hydroxy-5-methyl-4-isoazole-propionic acid) and N-methyl-D-aspartate (NMDA) receptors [30]. AMPA receptors activate and deactivate within a few milliseconds of presynaptic glutamate release, whereas the open probability of NMDA receptors typically reaches a peak after 20–30 ms, and decays over hundreds of milliseconds [31, 32]. This synaptic input is modelled by (8.7).

$$\varepsilon_{ij}^{PC\leftarrow PC}(t) = V_{EPSP}(t) = V_{AMPA}(t) + V_{NMDA}(t) \text{ .} \tag{8.7}$$

$V_{AMPA}(t)$ is the AMPA potential governed in (8.8),

$$V_{AMPA}(t) = \tag{8.8}$$

$$\frac{A_{AMPA}}{aN}\left(\frac{t - t_j^{(f)} - t_{delay}}{\tau_{AMPA}}\right)\exp\left(1 - \frac{t - t_j^{(f)} - t_{delay}}{\tau_{AMPA}}\right) ,$$

where $A_{AMPA} = 23.1\,\text{mV}$, $\tau_{AMPA} = 1.5\,\text{ms}$ (see [1] for approximate value), and $V_{NMDA}(t)$ is the NMDA potential governed in (8.9),

$$V_{NMDA}(t) = \frac{A_{NMDA}}{aN} \cdot \tag{8.9}$$

$$\exp\left(-\frac{t - t_j^{(f)} - t_{delay}}{\tau_{NMDA,f}}\right)\left(1 - \exp\left(-\frac{t - t_j^{(f)} - t_{delay}}{\tau_{NMDA,r}}\right)\right),$$

where N denotes the number of pyramidal cells or interneurons in the network, and the delay in the recurrent feedback is $t_{delay} = 0.5\,\text{ms}$. The constant $a = M/N$ denotes the sparseness of pyramidal cells for information coding with M denoting the number of cells representing a memory pattern. In this chapter, the slow NMDA-mediated component is ignored ($A_{NMDA} = 0\,\text{mV}$). Slow NMDA channels have a time constant that spans several gamma cycles which will allow synaptic connections to be formed between pyramidal cells that represent different memories. As a consequence, slow NDMA channels enable the storage of heteroassociative sequence information in long-term memory [5], in which NMDA receptors are responsible for learning novel paired associations [33]. Figure 8.2b illustrates the EPSP kernel function that describes the recurrent collaterals.

For synaptic transmission from an interneuron j to pyramidal cell i, the response kernel $\varepsilon_{ij}^{PC \leftarrow IN}$ denotes an inhibitory postsynaptic potential (IPSP) V_{IPSP}. V_{IPSP} represents the net GABAergic inhibitory feedback to the pyramidal cells from one interneuron. Recent evidence from cholinergically induced gamma-frequency network oscillations in vitro, shows that perisomatic-targeting GABAergic interneurones provide prominent rhythmic inhibition in pyramidal cells, and that these interneurones are synchronized by recurrent excitation [34, 35]. This excitatory-inhibitory feedback loop is sufficient to generate the intrahippocampal gamma-frequency oscillations [36]. The recruitment of this recurrent excitatoryinhibitory feedback loop during hippocampal gamma oscillations suggests that local gamma oscillations not only control when, but also how many and which pyramidal cells will fire during each gamma cycle [35]. In addition, interneurons fire at gamma frequency on positive portion of the theta oscillation [37]. This inhibitory effect is modelled in (8.10) by assuming the interneurons fire in synchrony by recurrent excitation of the pyramidal cells.

$$\varepsilon_{ij}^{PC \leftarrow IN}(t) = V_{IPSP}(t) \tag{8.10}$$

$$= \frac{A_{GABA}}{aN} \left(\frac{t - t_j^{(f)}}{\tau_{GABA}} \right) \exp\left(1 - \frac{t - t_j^{(f)}}{\tau_{GABA}} \right),$$

where $A_{GABA} = -5.94\,\text{mV}$, $t_j^{(f)}$ refers to the spike time of interneuron j, $\tau_{GABA} = 4\,\text{ms}$ (see [38] for approximate value). Figure 8.2c illustrates the feedback inhibition from one interneuron.

8.2.3   SRM-Based Interneuron

Action potentials fired by CA3 pyramidal cells could initiate inhibitory postsynaptic potentials (IPSPs) in nearby pyramidal cells [38]. One pyramidal cell could activate several disynaptic inhibitory pathways terminating on another pyramidal cell. This is suggestive of a diverse excitation of inhibitory cells to ensure recurrent inhibition is sufficiently widespread, rapid and potent to control the spread of activity by recurrent excitatory connections between CA3 pyramidal cells.

For simplicity, the response kernel $\varepsilon_{ij}^{IN \leftarrow PC}$ of interneuron i to presynaptic spikes from pyramidal cell j is modelled by (8.11). The synaptic transmission $w_{ij}^{IN \leftarrow PC}$ from pyramidal cell j to interneuron i is assumed with unit weight. This set-up simply serves as a signal from a pyramidal cell to initiate IPSP to other pyramidal cells via the inhibitory feedback from the interneurons.

$$\varepsilon_{ij}^{IN \leftarrow PC}(t - t_j^{(f)}) = \tag{8.11}$$

$$\begin{cases} 4 \text{ mV (see [38])} & \text{if } (t - t_j^{(f)}) = 0 , \\ 0 \text{ mV} & \text{if } (t - t_j^{(f)}) \neq 0 . \end{cases}$$

Thus, the firing threshold of the interneurons is assumed $V_{thres}^{IN} = 4\,\text{mV}$ (see [38] for approximate value). The interneuron is also assumed with no refractory ($\eta(s) = 0$) and no other external signal ($h^{ext}(t) = 0$) to simplify the kinetics of the interneurons.

8.3   Convergence of Synaptic Weight

Auto-associative LTM is formed by storing the associative information in synapses between pyramidal cells at the recurrent collaterals [1]. Jensen et al. proposed a synaptic modification methodology at the recurrent collaterals as follows in (8.12).

$$\frac{dw_{ij}^{PC \leftarrow PC}}{dt} = \left(\frac{i_{post}(t - t_j^{(f)}) \cdot b_{glu}(t - t_i^{(f)} - t_{delay})}{\tau_{pp}} \right)(1 - w_{ij}^{PC \leftarrow PC}) \tag{8.12}$$

$$+ \left(\frac{i_{post}(t - t_j^{(f)})}{\tau_{npp}} + \frac{b_{glu}(t - t_i^{(f)} - t_{delay})}{\tau_{pnp}} \right)(0 - w_{ij}^{PC \leftarrow PC})$$

where $i_{post}(.)$ denotes the postsynaptic depolarization that is attributed to back-propagating action potentials or other dentritic depolarizing events that occur after the spike initiation in the somatic region and is defined as

$$i_{post}(s) = \frac{s}{\tau_{post}} \exp\left(1 - \frac{s}{\tau_{post}}\right) .$$

$b_{glu}(.)$ models the time course of the glutamate bound to the NMDA receptors and is defined as

$$b_{glu}(s) = \exp\left(\frac{s}{\tau_{NMDA,f}}\right)\left(1 - \exp\left(-\frac{s}{\tau_{NMDA,r}}\right)\right) .$$

Equation (8.12) is a first order linear differential equation and can be rewritten as

$$\frac{dw_{ij}^{PC \leftarrow PC}}{dt} = -(A(t) + B(t)) w_{ij}^{PC \leftarrow PC} + A(t) ,$$

where

$$A(t) = \frac{i_{post}(t - t_j^{(f)}) \cdot b_{glu}(t - t_i^{(f)} - t_{delay})}{\tau_{pp}},$$

and

$$B(t) = \frac{i_{post}(t - t_j^{(f)})}{\tau_{npp}} + \frac{b_{glu}(t - t_i^{(f)} - t_{delay})}{\tau_{pnp}}.$$

It is easy to see that the synaptic dynamics is stable and it will converge to (8.13).

$$w_{ij}^{*PC\leftarrow PC} = \exp\left(-\int_0^{+\infty} (A(s) + B(s))\, ds\right) \qquad (8.13)$$
$$\cdot \left(w_{ij}^{PC\leftarrow PC}(0) + \int_0^{+\infty} A(s) \exp\left(\int_0^s (A(\xi) + B(\xi))\, d\xi\right) ds\right)$$

Equation (8.13) is not mathematically solvable. In order to estimate the limit of the synaptic weight between any two cells encoding the same memory, the integration terms in (8.13) are pre-computed at discrete time step of 0.1 ms. This is possible since the kernel functions $A(t)$ and $B(t)$ are known. The synaptic weight modification method proposed by Jensen et al. would only adjust synaptic weight $w_{ij}^{PC\leftarrow PC}$ between cell i and cell j encoding the same memory item to approach 0.534. Figure 8.3a shows weight $w_{ij}^{PC\leftarrow PC}$ converges to 0.534 from different initial values. Figure 8.3b illustrates the synaptic weight converges to 0.431 when the kinetics of fast NMDA receptors are modified. Here, the parameters $\tau_{NMDA,f}$ and $\tau_{NMDA,r}$ that characterize the kinetics of the NMDA receptors are doubled and are respectively 14 and 2 ms.

The two examples illustrate synaptic weight modification rule is stable. However, changing the parameters of the learning rule resulted in synaptic weight converging to another value. Should a different number of cells fired, a static synaptic weight modification rule will certainly fail. The following two sections describe the conditions under which this learning rule will fail.

8.4 Maximum Synaptic Weight to Prevent Early Activation

The network incorporates a STM network where each memory item is repeated within a gamma subcycle of every theta cycle. The pyramidal cells are reactivated by the ADP that is intrinsic to each cell. If the cells that encode a memory item were to fire in another gamma cycle within the same theta cycle, the firing of these cells will potentially corrupt and interfere with the memory item in that gamma cycle. The

Fig. 8.3 Convergence of
synaptic strength between
synchronously-fired
pyramidal cells. **a** LTM
synaptic modification using
the original parameters in
[1]. **b** Original LTM synaptic
modification with $\tau_{NMDA,f} =$
14 ms and $\tau_{NMDA,r} = 2$ ms

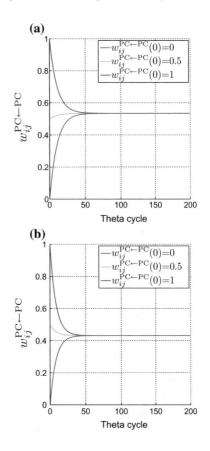

firing of the cells that encode distinct memory items would thus be "link up" by the
LTM synaptic modification proposed by Jensen et al. [1].

In order to prevent auto-associative LTM corruption, cells must be prevented
from firing in other cycles other than in its designated gamma cycle. The membrane
potential of the pyramidal cells is governed by (8.3). After a pyramidal cell has
fired, its membrane potential will be reset. A pyramidal cell will only fire after its
membrane potential $u_i^{PC}(t)$ exceeds V_{thres}^{PC} which is governed by (8.14).

$$u_i^{PC}(t) = \eta_i^{PC}\left(t - \hat{t}_i\right) + \sum_{j \in \Gamma_i} w_{ij}\varepsilon_{ij}\left(t - \hat{t}_j\right) + h^{ext}(t) \qquad (8.14)$$

$$\geq V_{thres}^{PC}.$$

The pyramidal cells encoding a memory item will fire much earlier when the
synaptic weight $w_{ij}^{PC \leftarrow PC}$ between cell i and cell j of the recurrent collaterals is high
enough to trigger the same set of cells to fire in the next gamma cycle. In order
to prevent cells from reactivating within the same theta cycle and corrupting other

memory items, the weight of the synapses in the recurrent collaterals that encode the LTM for the same memory item must be upper bounded. The maximum synaptic weight w_{max} between cell i and cell j of the recurrent collaterals is determined by solving (8.15).

$$u_i^{PC}(t) = \eta_i^{PC}\left(t - \hat{t}_i\right) + \sum_{j \in \Gamma_i} w_{ij}\varepsilon_{ij}\left(t - \hat{t}_j\right) + h^{ext}(t) \qquad (8.15)$$
$$< V_{thres}^{PC} .$$

A memory item is encoded by a set of M pyramidal cells associated by their synaptic weights ($w_{ij}^{PC \leftarrow PC}$ between cell i and cell j of the recurrent collaterals). Ideally, $w_{ij}^{PC \leftarrow PC} \leq w_{max}$ for $i \neq j$, $|i - j| < M$ and $pM < i, j \leq (p + 1)M$ where $p \in \mathbb{Z}_P$ ($\mathbb{Z}_P$ = finite set of integers modulo P, where P is the number of memory patterns), and $w_{ij}^{PC \leftarrow PC} = 0$ otherwise. Given that M pyramidal cells fire for each memory item, M interneurons will also fire. Thus (8.15) can further be simplified to solve for w_{max}. In order to prevent the set of M pyramidal cells from early reactivation, the membrane potential of pyramidal cell i must satisfy (8.16).

$$u_i^{PC}(t) = \eta_i^{PC}\left(t - \hat{t}_i\right) + (M - 1)w_{max}V_{EPSP}(t) \qquad (8.16)$$
$$+ M V_{IPSP}(t) + h^{ext}(t) < V_{thres}^{PC} .$$

Thus, the maximum synaptic weight w_{max} between cell i and j of the recurrent collaterals is determined by (8.17).

$$w_{max} < \frac{V_{thres}^{PC} - h^{ext}(t) - \eta_i^{PC}\left(t - \hat{t}_i\right) - M V_{IPSP}(t)}{(M - 1)V_{EPSP}(t)} . \qquad (8.17)$$

The maximum weight w_{max} is a dynamic synaptic weight upper bound defined by the different kernel functions with time varying characteristics, as well as, the time at which pyramidal cells fire. After a group of pyramidal cells has fired, a cell i belonging to the same group will not re-fire as long as its membrane potential u_i^{PC} does not exceed its firing threshold V_{thres}^{PC}. If synaptic weight of synapses between cells encoding the same memory pattern were to exceed the upper bound, the membrane potential of the cell will exceed the firing threshold and the cell will fire.

8.5 Pattern Completion of Auto-Associative Memory

Pyramidal cells encoding the same memory item are auto-associated by the synapses within the recurrent collaterals. A memory item can be recalled when a smaller set of cells are fired [19]. However, a minimum number of cells encoding the same memory must fire to trigger the other cells and retrieve the memory item.

The membrane potential u_i^{PC} of a pyramidal cell i is governed by (8.3). A minimum number of cells is needed to fire other pyramidal cells of the same auto-associative memory. It is required that the membrane potential of these other cells exceed V_{thres}^{PC}. The firing of the other pyramidal cells is governed by the summation of all the V_{EPSP} due to firing of the minimum number of pyramidal cells encoding the same memory item and the inhibitory feedback from interneurons due to the firing of these pyramidal cells.

Let $M^{(f)}$ denotes the minimum number of pyramidal cells needed to fire in order to trigger the other cells encoding the same memory item, and retrieve the stored memory item. The membrane potential u_i^{PC} of other cell i within the set of cells encoding the same memory item must exceed V_{thres}^{PC} is governed by (8.18).

$$u_i^{PC}(t) = \sum_j^{M^{(f)}} w_{ij}^{PC \leftarrow PC} V_{EPSP}(t) + \sum_j^{M^{(f)}} V_{IPSP}(t) + h^{ext}(t) \quad (8.18)$$

$$\geq V_{thres}^{PC}; \quad pM < i, j \leq (p+1)M; \quad p \in \mathbb{Z}_P .$$

Thus, the synaptic weight $w_{ij}^{PC \leftarrow PC}$ between cell i and cell j of the recurrent collaterals must be lower bounded to ensure that cells that encode the same memory item are triggered. If pyramidal cells encoding the same memory item are connected with synapses in the recurrent collaterals with weight w_{min} ($w_{ij}^{PC \leftarrow PC} = w_{min}; i \neq j$; $|i - j| < M; pM < i, j \leq (p+1)M; p \in \mathbb{Z}_P$), the minimum synaptic weight w_{min} between cell i and cell j is determined by (8.19).

$$w_{min} = \frac{V_{thres}^{PC} - h^{ext}(t) - \sum_j^{M^{(f)}} V_{IPSP_{ij}}(t)}{\sum_j^{M^{(f)}} V_{EPSP}(t)} . \quad (8.19)$$

Similar to the synaptic weight upper bound, the lower bound w_{min} is also time varying in nature as it depends on the different kernel functions. The dynamic synaptic weight lower bound w_{min} is also calculated with respects to the last firing time of the $M^{(f)}$ presynaptic pyramidal cells. Here, it is assumed that the $M^{(f)}$ presynaptic pyramidal cells fire together. The firings of these pyramidal cells act as a cue to trigger the other cells which together encode the same memory item. Pattern completion is successfully when these "missing cells" are triggered.

8.6 Discussion

Figure 8.4 illustrates the process of pattern completion for $M = 5$ when 80% of every memory pattern are presented to the network. The figure shows the result for simulation with 0.1 ms time step. The weight of the recurrent synapses between cells encoding the same memory patterns are set to 0.534 (i.e. $w_{ij}^{PC \leftarrow PC} = 0.534$ for $pM < i, j \leq (p+1)M; p \in \mathbb{Z}_7; i \neq j$). This is the weight recurrent synapses will

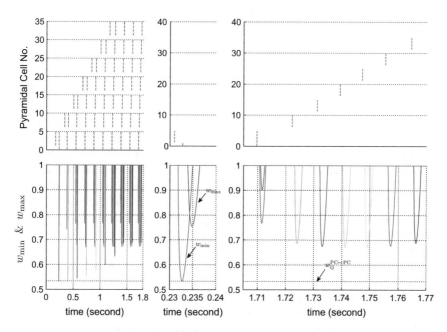

Fig. 8.4 Pattern completion when $w_{ij}^{PC \leftarrow PC} = 0.534$ for $pM < i, j \le (p+1)M$; $p \in \mathbb{Z}_7$ and $M = 5$. *Left* to *right* column: firing time of pyramidal cells on *top row* and corresponding minimum and maximum synaptic weight for each group of cells representing the same memory item at different time scale. *Top row* firing time of each pyramidal cell. *Short line* denotes the firing of the corresponding pyramidal cell. *Red lines* indicate cell firings are unsynchronized or pattern completion is unsuccessful. *Blue lines* indicate pattern completion is successful. Pattern completion can only be carried out for memory pattern 1 (cell 1 to cell 5 fire in synchrony). *Bottom row* lines with same colour represent the maximum and minimum synaptic weight for synapses between each subset of cells encoding the same memory. *Bold lines* represent the maximum weight, and regular lines represents the minimum recurrent synaptic weight

converge to using the synaptic modification method presented in [1]. The top row of Fig. 8.4 shows the firing time of each pyramidal cell with increasing time resolution from left to right. The top-left plot shows the firing time of pyramidal cells from time $t = 0$ s to $t = 1.8$ s, the top-middle plot shows the firing time of pyramidal cells from $t = 0.23$ s to $t = 0.24$ s, and the top-right plot shows the firing time of pyramidal cells from $t = 1.7$ s to $t = 1.77$ s. The bottom row of Fig. 8.4 shows the corresponding minimum and maximum synaptic weights between cells in each group encoding the same memory item at different time. Each bold line in these plots represents the maximum synaptic weight for synapses between cells encoding one memory item. Each regular line represents the minimum synaptic weight for synapses between cells encoding a memory item. Red lines in the lower plots represent the weight bounds for synapses between cells encoding the first memory item. Green lines represent the synaptic bounds for memory item 2, and so forth for memory item 3 to memory item 7. Synaptic weight is capped between [0, 1]. The bottom-left plot shows the dynamic lower and upper weights for synapses between cells encoding the same memory for $t = 0$s to $t = 1.8$s. This plot corresponds to the cell firing time in the top-left plot. The bottom-middle and bottom-right plots respectively correspond to the firing time of pyramidal cells in the top-middle and top-right plots.

The first partial memory pattern is injected into cell 2 to cell 5 ($M^{(f)} = 4$) at time $t = 166.6$ ms. Lines are drawn at time $t = 166.6$ ms to indicate the firing of cell 2 to cell 5 as shown in top-left plot. At this time, pattern completion is still unsuccessful for memory item 1 as cell 1 has not been fired. These lines are drawn in red to indicate incomplete recall of this memory item. ADP mechanism intrinsic to every pyramidal cell kicks in and couples with theta oscillation, the membrane potentials of cell 2 to cell 5 subsequently exceed the firing threshold. This is shown in the top-middle plot where cell 2 to cell 5 are synchronously activated at time $t = 231.0$ ms. Similarly red lines are drawn at time $t = 231.0$ ms in the top-middle plot of Fig. 8.4 to indicate the firing of these cells. At this time, completion of memory pattern 1 is still unsuccessful since cell 1 has not been fired.

Cell 1 is associated with cell 2 to cell 5 as these cells together encode memory item 1. The activation of cell 2 to cell 5 subsequently triggers cell 1 to fire by the excitatory postsynaptic response to spikes from cell 2 to cell 5 via the recurrent collaterals. A red line is drawn at time $t = 232.8$ ms to indicate the firing of cell 1. ADP and theta oscillation together causes cell 2 to cell 5 to fire in the next theta cycle at time $t = 377.8$ ms, and cell 1 to fire at $t = 378.4$ ms. At this time, pattern completion is still unsuccessful. ADP of every cell is activated and together with theta oscillation, these cells fire again in the next theta cycle. However, pattern completion of memory item 1 is successful at this cycle. All cells of memory item 1 are synchronously fired at $t = 543.2$ ms. Blue lines are drawn to indicate the successful completion of memory item. At $t = 543.2$ ms, cell 1 to cell 5 fire in synchrony.

Pattern completion for memory item 1 is possible because the synaptic weight of 0.534 exceeds the minimum synaptic weight for synapses between cell 1 and the other cells (cell 2 to cell 5) encoding memory item 1. The minimum synaptic weight to allow cell 2 to cell 5 to trigger cell 1 is shown in the bottom-middle plot. The plot shows a valley-like minimum synaptic weight. At $t = 232.8$ ms, the minimum synaptic weight decreases below 0.534. At this point, the potential of cell 1 exceeds its firing threshold and cell 1 fires (see top-middle plot). After cell 1 fires, the maximum synaptic weight for synapses among this group of cells for memory item 1 decreases. Should the weight of these synapses exceeds 0.7542 (the smallest value of the maximum synaptic weight curve represented by red bold line in bottom-middle plot), cell 1 to cell 5 would be fired. Since weight of 0.534 is below this mark, these cells did not fire prematurely.

The second partial memory pattern is injected into cell 7 to cell 10 at time $t = 333.3$ ms. Similarly, ADP and theta oscillation causes cell 7 to cell 10 to re-fire at $t = 399.7$ ms. Thereafter, the minimum synaptic weight that is able to trigger cell 6 decreases due to the increasing excitatory postsynaptic potential at cell 6 in response to spikes from cell 7 to cell 10. However, the weight of 0.534 is below the minimum allowable weight to trigger cell 6. Thus, pattern completion for memory item 2 is unsuccessful. This phenomenon also happens to pattern completion of memory item 3 to memory item 7. Generally, a higher weight is required to ensure the successful recall of subsequent memory item. This is due to stronger inhibition from interneurons accumulated due to the activation of other pyramidal cells. Figure 8.4 shows that pattern completion can only be carried out for memory pattern 1. For the other patterns, the recurrent synaptic weights are not high enough to trigger the other missing cell that also encodes the same pattern.

Figure 8.5a illustrates the mechanism of STM network when the recurrent synaptic weight exceeds the maximum allowable value. Figure 8.5b shows a portion of the simulation result. There are periods of time within the positive cycle of the theta oscillation when pyramidal cell continues to re-fire. For example, the continual activation of cell 1 to cell 5 within the first theta cycle. This phenomenon is coupled by the high feedback of EPSP in response to spikes from other pyramidal cells and low IPSP in response to spikes from the interneurons. The high excitatory response is due to the high recurrent synaptic weight factoring onto the excitatory responses to spikes from associated pyramidal cells. The low inhibitory response is due to the use of the short-term variant of SRM model that relies only on the most recent spike to compute the inhibitory effect. Since the spiking of cell is maintained by the slow-ramping ADP of the refractory kernel, this phenomenon will repeat for every theta cycle. However the continual burst of activation is terminated when there is sufficient inhibition. For example, the continuous firings of cells for memory pattern 1 are terminated after new input from cell 7 to cell 10 produces sufficient inhibition. However, cells encoding memory pattern 1 subsequently fire in synchrony with cells encoding memory pattern 2, which results in the merger of these memory patterns in the STM network. Memory pattern 3 and 4 are also subsequently merged in the STM network. Should LTM synaptic modification be carried out, auto-associative LTM would be adapted and cells encoding different memory patterns would be associated. This mechanism could help explain the asymmetric distribution of spike density in theta cycle [16].

In general, the future state of the network is dependent on the state at which the network is presently in. For example, if a different synaptic weight were used, the minimum and maximum synaptic weights would differ since both are dependent on the amount of excitatory response to spikes from presynaptic pyramidal cells (see (8.17) and (8.19)). The activation of one STM pattern would produce excitatory response to other cells which would affect the activation of STM pattern in subsequent gamma subcycles.

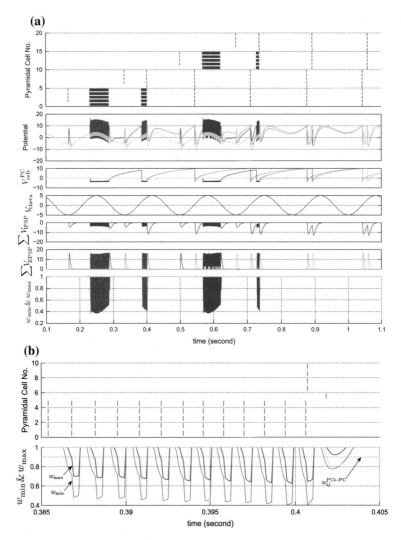

Fig. 8.5 STM mechanism when $w_{ij}^{PC \leftarrow PC} = 0.9$ for $pM < i, j \leq (p + 1)M$; $p \in \mathbb{Z}_4$ and $M = 5$. **a** *Top row* firing time of each pyramidal cell. Short line indicates the firing of corresponding pyramidal cell. *Red line* indicates cell firings are unsynchronized within the group of cells encoding the same memory pattern or pattern completion is unsuccessful. *Second row red, green, blue* and *cyan lines* respectively represent the membrane potential of cell 1, cell 6, cell 11 and cell 16. *Third row red, green, blue* and *cyan lines* respectively represent the refractory response of cell 1, cell 6, cell 11 and cell 16. *Fourth row* theta oscillation. *Fifth row* the total IPSP received by pyramidal cells. *Sixth row red, green, blue* and *cyan lines* respectively represent the EPSP received by cell 1, cell 6, cell 11 and cell 16. *Last row* minimum and maximum synaptic weight. *Red, green, blue* and *cyan* coloured *bold lines* respectively represent the maximum synaptic weight for synapses between cells encoding memory item 1, 2, 3 and 4. *Red, green, blue* and *cyan* coloured regular lines respectively represent the minimum synaptic weight for synapses between cells encoding memory item 1, 2, 3, and 4. Here, the minimum and maximum synaptic weights are computed without merger of STM patterns. **b** Enlarged plot of Fig. 8.5a from time 0.385–0.405 s

8.7 Conclusion

The hypothesis of associative memory storage and recall in the different components of the hippocampus is currently not possible to be evaluated by direct tests. This is because it is technically not possible to directly generate specific memory patterns of neural activity to demonstrate either storage or subsequent recall. The hypothesis that such a mechanism of associative memory formation and recall does occur is based on the suitability of the neural network architecture and experiments of Hebbian induction of long-lasting changes in synaptic strength at relevant synapses on tissue slice [39, 40].

Here, computational model is used to assess the auto-associative memory storage and recall abilities of the hippocampal CA3 subsystem. The computational model presented here is based on Jensen et al.'s recurrent spiking neural network as model of subregion of hippocampal CA3 [1]; sufficient recurrent connectivity for auto-associative memory function may be restricted to subregion of CA3a [41]. Specifically, the computational model is analyzed to present the specific synaptic conditions that allow successful storage of different memory patterns and recall of previously stored patterns. This model has similarities between variety of models that consist of different levels of biological realism in the investigation of auto-associative memory function [1, 3, 14, 15, 41–46].

The present computational model holds memory patterns within its auto-associative network in the recurrent collaterals through STDP learning. Successful retrieval of correctly stored patterns is achievable if the synaptic conditions for the recurrent collaterals are satisfied. By connecting this auto-associative network with another similar recurrent network that functions as hetero-associative memory network [5, 19], it is possible to encode episodic memories and hold these data sequence for later retrieval within its auto- and hetero-associative networks. This combined network can effectively perform both of these functions within the same learning task, a mechanism suggested by [10]. This dual operating modes are possible with theta-rhythmic oscillation that naturally parses the CA3 function into encoding and retrieval cycles, which are also demonstrated in the model by [44]. Anatomical findings of the recurrent collaterals in CA3 subregions and physiological findings of STDP in dendrites also suggest the dual auto- and hetero-associative functionalities of CA3 [47].

Within the auto-associative memory model, it was also illustrated that feedback inhibition from the interneurons needs to accurately reflect pyramidal cell activity [48]. This requirement is similar to the working principle of some network models in [3, 42, 43]. The presented network is similar to the "pseudo-inhibition" models by [3, 43]. Each pyramidal cell provides an inhibitory connection via an interneuron onto all other pyramidal cells. However, a low inhibitory response due to the use of STM-variant of the SRM spiking neuron model in current network is unable to prevent the bursting of pyramidal cells. Thus, the firing rate of the interneurons should proportionally reflect the neural activities of the pyramidal cell population.

The results of this chapter are based on the generic SRM neuron model. SRM neuron model is able to describe other threshold-fire models like integrate-and-fire (I&F) and approximate the Hodgkin–Huxley conductance-based neuron model [14, 15]. Thus, the results of this chapter are extensible to spiking recurrent network that utilizes threshold-fire spiking neuron [14, 15]. In addition, other aspects of the network under different conditions can be analysed using the generic mathematical formulation (Eq. 8.3). This work finds the synaptic bounds of a recurrent collateral network for auto-associative memory formation. The same method can be applied onto spiking network that realizes hetero-associative memory formation.

The analysis on the bounds of the recurrent network reveals addition evidence for rejecting the view that STM and LTM are two separate entities. Through the interaction between LTM and STM network, evenly distribution spike density in theta cycle can become asymmetric, similar to the experimental result in [16]. Persistent burst of one STM pattern is merged with subsequent memory pattern. This is accomplished in two steps. Firstly, sufficient inhibition terminates the continual burst of the first group of cells via the activation of another subset of cells. Secondly, the after-depolarization (ADP) intrinsic to cells causes the two groups of cells to subsequently fire in synchrony.

The original work by [1] has been significant to the understanding of the recurrent spiking neural network dynamics. The present work relies on their model to illustrate the following points. Firstly, it is important that long-term memory (LTM) synaptic modification is coordinated to maintain network stability [17]. If a different number of cells other than the designed number of pyramidal cells were to be used to encode the memory patterns, the static LTM synaptic modification technique that updates synapses without the consideration of this factor will certainly not able to adapt synaptic weight within the new synaptic weight lower and upper bounds. Secondly, the analysis is applicable to other spiking network utilizing threshold-fire spiking neuron model by the use of a generic mathematical formulation (spiking response model [14, 15]). Thirdly, asymmetric distribution of spike density in theta cycle [16] can be explained by the the the merger of STM patterns through LTM and STM networks interaction.

References

1. Jensen, O., Idiart, M., Lisman, J.E.: Physiologically realistic formation of autoassociative memory in networks with theta/gamma oscillations: role of fast nmda channels. Learn. Mem. **3**(2–3), 243–256 (1996)
2. Rolls, E.: Computational models of hippocampal functions. Learning and memory: a comprehensive reference, pp. 641–665 (2008)
3. Cutsuridis, V., Wennekers, T.: Hippocampus, microcircuits and associative memory. Neural Netw. **22**(8), 1120–1128 (2009)
4. Rolls, E.T.: A computational theory of episodic memory formation in the hippocampus. Behav. Brain Res. **215**(2), 180–196 (2010)

5. Jensen, O., Lisman, J.E.: Theta/gamma networks with slow nmda channels learn sequences and encode episodic memory: role of nmda channels in recall. Learn. Mem. **3**(2–3), 264–278 (1996)

6. Bragin, A., Jandó, G., Nádasdy, Z., Hetke, J., Wise, K., Buzsáki, G.: Gamma (40–100 hz) oscillation in the hippocampus of the behaving rat. J. Neurosci. **15**(1), 47–60 (1995)

7. Vanderwolf, C.H.: Hippocampal electrical activity and voluntary movement in the rat. Electroencephalogr. Clin. Neurophysiol. **26**(4), 407–418 (1969)

8. Cantero, J.L., Atienza, M., Stickgold, R., Kahana, M.J., Madsen, J.R., Kocsis, B.: Sleep-dependent θ oscillations in the human hippocampus and neocortex. J. Neurosci. **23**(34), 10897–10903 (2003)

9. Hughes, J.R.: Gamma, fast, and ultrafast waves of the brain: their relationships with epilepsy and behavior. Epilepsy Behav. **13**(1), 25–31 (2008)

10. Hasselmo, M.E., Bodelón, C., Wyble, B.P.: A proposed function for hippocampal theta rhythm: separate phases of encoding and retrieval enhance reversal of prior learning. Neural Comput. **14**(4), 793–817 (2002)

11. Lisman, J.E., Idiart, M.A.: Storage of $7 +/- 2$ short-term memories in oscillatory subcycles. Science **267**(5203), 1512–1515 (1995)

12. Wagatsuma, H., Yamaguchi, Y.: Neural dynamics of the cognitive map in the hippocampus. Cognit. Neurodyn. **1**(2), 119–141 (2007)

13. Yamaguchi, Y., Sato, N., Wagatsuma, H., Wu, Z., Molter, C., Aota, Y.: A unified view of theta-phase coding in the entorhinal-hippocampal system. Curr. Opin. Neurobiol. **17**(2), 197–204 (2007)

14. Maass, W., Bishop, C.M.: Pulsed Neural Networks. MIT Press, Cambridge (2001)

15. Gerstner, W., Kistler, W.M.: Spiking Neuron Models: Single Neurons, Populations, Plasticity, 1st edn. Cambridge University Press, Cambridge (2002)

16. Koene, R.A., Hasselmo, M.E.: First-in-first-out item replacement in a model of short-term memory based on persistent spiking. Cereb. Cortex **17**(8), 1766–1781 (2007)

17. Abbott, L.F., Nelson, S.B.: Synaptic plasticity: taming the beast. Nature Neurosci. **3**, 1178–1183 (2000)

18. Tan, C.H., Cheu, E.Y., Hu, J., Yu, Q., Tang, H.: Associative memory model of hippocampus ca3 using spike response neurons. In: Processing of the Neural Information, pp. 493–500. Springer, Heidelberg (2011)

19. Jensen, O., Lisman, J.E.: Novel lists of $7+/-2$ known items can be reliably stored in an oscillatory short-term memory network: interaction with long-term memory. Learn. Mem. **3**(2–3), 257–263 (1996)

20. Jensen, M.S., Azouz, R., Yaari, Y.: Spike after-depolarization and burst generation in adult rat hippocampal ca1 pyramidal cells. J. Physiol. **492**, 199–210 (1996)

21. Storm, J.F.: An after-hyperpolarization of medium duration in rat hippocampal pyramidal cells. J. Physiol. **409**(1), 171–190 (1989)

22. Park, J.Y., Remy, S., Varela, J., Cooper, D.C., Chung, S., Kang, H.W., Lee, J.H., Spruston, N.: A post-burst afterdepolarization is mediated by group i metabotropic glutamate receptor-dependent upregulation of cav2. 3 r-type calcium channels in ca1 pyramidal neurons. PLoS Biol. **8**(11), e1000534 (2010)

23. Araneda, R., Andrade, R.: 5-hydroxytryptamine 2 and 5-hydroxytryptamine 1a receptors mediate opposing responses on membrane excitability in rat association cortex. Neuroscience **40**(2), 399–412 (1991)

24. Alonso, A., Gaztelu, J., Bun, W., Garcia-Austt, E., et al.: Cross-correlation analysis of septo-hippocampal neurons during$\equiv$-rhythm. Brain Res. **413**(1), 135–146 (1987)

25. Rutishauser, U., Ross, I.B., Mamelak, A.N., Schuman, E.M.: Human memory strength is predicted by theta-frequency phase-locking of single neurons. Nature **464**(7290), 903–907 (2010)

26. Skaggs, W.E., McNaughton, B.L.: Theta phase precession in hippocampal. Hippocampus **6**, 149–172 (1996)

27. Yamaguchi, Y., Aota, Y., McNaughton, B.L., Lipa, P.: Bimodality of theta phase precession in hippocampal place cells in freely running rats. J. Neurophysiol. **87**(6), 2629–2642 (2002)

28. Wagatsuma, H., Yamaguchi, Y.: Cognitive map formation through sequence encoding by theta phase precession. Neural Comput. **16**(12), 2665–2697 (2004)
29. Mizuseki, K., Sirota, A., Pastalkova, E., Buzsáki, G.: Theta oscillations provide temporal windows for local circuit computation in the entorhinal-hippocampal loop. Neuron **64**(2), 267–280 (2009)
30. Ozawa, S., Kamiya, H., Tsuzuki, K.: Glutamate receptors in the mammalian central nervous system. Prog. Neurobiol. **54**(5), 581–618 (1998)
31. Forsythe, I.D., Westbrook, G.L.: Slow excitatory postsynaptic currents mediated by n-methyl-d-aspartate receptors on cultured mouse central neurones. J. Physiol. **396**(1), 515–533 (1988)
32. Stern, P., Edwards, F.A., Sakmann, B.: Fast and slow components of unitary epscs on stellate cells elicited by focal stimulation in slices of rat visual cortex. J. Physiol. **449**(1), 247–278 (1992)
33. Rajji, T., Chapman, D., Eichenbaum, H., Greene, R.: The role of ca3 hippocampal nmda receptors in paired associate learning. J. Neurosci. **26**(3), 908–915 (2006)
34. Mann, E.O., Radcliffe, C.A., Paulsen, O.: Hippocampal gamma-frequency oscillations: from interneurones to pyramidal cells, and back. J. physiol. **562**(1), 55–63 (2005)
35. Hájos, N., Paulsen, O.: Network mechanisms of gamma oscillations in the ca3 region of the hippocampus. Neural Netw. **22**(8), 1113–1119 (2009)
36. Caillard, O., Debanne, D.: Cell-specific contribution to gamma oscillations. J. Physiol. **588**(5), 751–751 (2010)
37. Sik, A., Penttonen, M., Ylinen, A., Buzsáki, G.: Hippocampal ca1 interneurons: an in vivo intracellular labeling study. J. Neurosci. **15**(10), 6651–6665 (1995)
38. Miles, R.: Synaptic excitation of inhibitory cells by single ca3 hippocampal pyramidal cells of the guinea-pig in vitro. J. Physiol. **428**(1), 61–77 (1990)
39. Bliss, T.: Synaptic plasticity in the hippocampus. Trends Neurosci. **2**, 42–45 (1979)
40. Neves, G., Cooke, S.F., Bliss, T.V.: Synaptic plasticity, memory and the hippocampus: a neural network approach to causality. Nature Rev. Neurosci. **9**(1), 65–75 (2008)
41. de Almeida, L., Idiart, M., Lisman, J.E.: Memory retrieval time and memory capacity of the ca3 network: role of gamma frequency oscillations. Learn. Mem. **14**(11), 795–806 (2007)
42. Marr, D.: Simple memory: a theory for archicortex. Philos. Trans. R. Soc. B Biol. Sci. **262**(841), 23–81 (1971)
43. Sommer, F.T., Wennekers, T.: Associative memory in networks of spiking neurons. Neural Netw. **14**(6), 825–834 (2001)
44. Kunec, S., Hasselmo, M.E., Kopell, N.: Encoding and retrieval in the ca3 region of the hippocampus: a model of theta-phase separation. J. Neurophysiol. **94**(1), 70–82 (2005)
45. Bush, D., Philippides, A., Husbands, P., O'Shea, M.: Dual coding with stdp in a spiking recurrent neural network model of the hippocampus. PLoS Comput. Biol. **6**(7), e1000839 (2010)
46. Cutsuridis, V., Cobb, S., Graham, B.P.: Encoding and retrieval in a model of the hippocampal ca1 microcircuit. Hippocampus **20**(3), 423–446 (2010)
47. Samura, T., Hattori, M., Ishizaki, S.: Autoassociative and heteroassociative hippocampal ca3 model based on location dependencies derived from anatomical and physiological findings. In: International Congress Series, vol. 1301, pp. 140–143. Elsevier (2007)
48. Hunter, R., Cobb, S., Graham, B.P.: Improving associative memory in a network of spiking neurons. In: Artificial neural networks-ICANN 2008, pp. 636–645. Springer (2008)